Praise for *Top-Bar Beekeeping*

"This is an excellent guide for hobby beekeepers who wish to keep bees using top-bar hives. Drawing on his more than thirty years of beekeeping experience in New Mexico, author Les Crowder describes in detail the special comb-management techniques that this low-cost, but relatively intensive, form of beekeeping requires. *Top-Bar Beekeeping* also provides an eloquent appeal for beekeepers to make care, respect, and reverence the foundation of their relationships with the bees."

—Thomas D. Seeley, Cornell University;
author of *Honeybee Democracy* and *The Wisdom of the Hive*

"Reading *Top-Bar Beekeeping* reminds me of the classes I took with Les Crowder several years ago. He's a man who truly knows whereof he speaks, who has the gift of communicating with his small friends, the bees, and sharing his understanding with us. . . . This is the one book on beekeeping that I will recommend to my permaculture students."

—Scott Pittman, director, Permaculture Institute USA

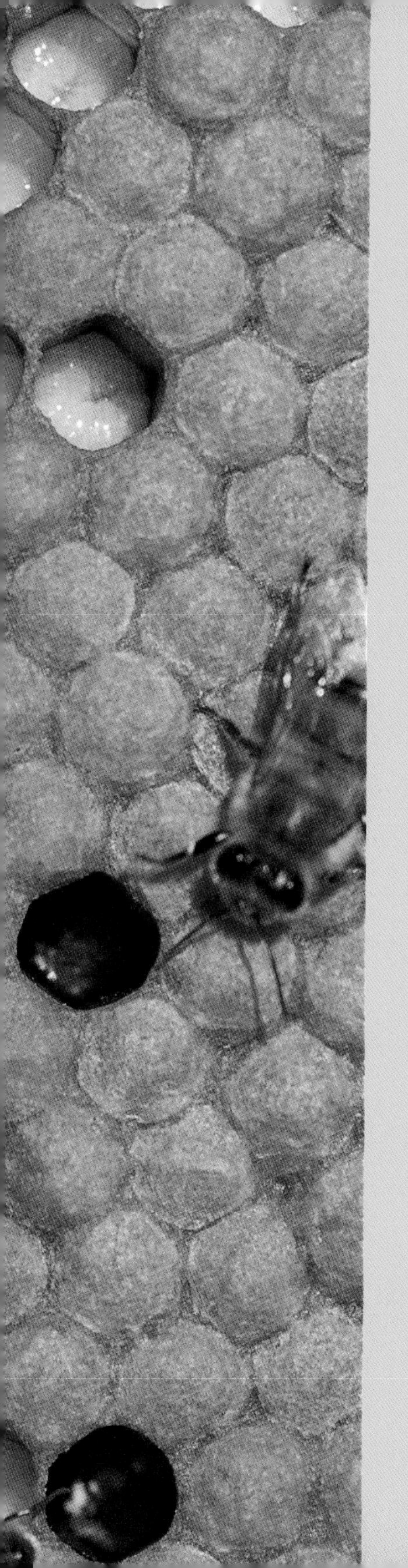

Top-Bar Beekeeping

ORGANIC PRACTICES FOR HONEYBEE HEALTH

LES CROWDER
and
HEATHER HARRELL

Chelsea Green Publishing
White River Junction, Vermont

Project Manager: Patricia Stone
Developmental Editor: Ben Watson
Copy Editor: Ellen Brownstein
Proofreader: Alice Colwell
Indexer: Shana Milkie
Designer: Melissa Jacobson

Printed in the United States of America
First printing August, 2012
10 9 8 7 6 5 4 3 2   13 14 15 16

Our Commitment to Green Publishing

Chelsea Green sees publishing as a tool for cultural change and ecological stewardship. We strive to align our book manufacturing practices with our editorial mission and to reduce the impact of our business enterprise in the environment. We print our books and catalogs on chlorine-free recycled paper, using vegetable-based inks whenever possible. This book may cost slightly more because it was printed on paper that contains recycled fiber, and we hope you'll agree that it's worth it. Chelsea Green is a member of the Green Press Initiative (www.greenpressinitiative.org), a nonprofit coalition of publishers, manufacturers, and authors working to protect the world's endangered forests and conserve natural resources. *Top-Bar Beekeeping* was printed on FSC®-certified paper supplied by CJK that contains at least 10% post-consumer recycled fiber.

Library of Congress Cataloging-in-Publication Data

Crowder, Les.
Top-bar beekeeping : organic practices for honeybee health / Les Crowder and Heather Harrell.
p. cm.
Includes bibliographical references and index.
ISBN 978-1-60358-461-6 (pbk.) — ISBN 978-1-60358-462-3 (ebook)
1. Bee culture. 2. Beehives. 3. Honeybee. I. Harrell, Heather. II. Title.

SF523.C858 2012
638'.1—dc23

2012020426

Chelsea Green Publishing
85 North Main Street, Suite 120
White River Junction, VT 05001
(802) 295-6300
www.chelseagreen.com

Contents

Note from Les Crowder

The ideas and methods detailed in this book have evolved over the course of nearly forty years. Many of the events and experiments took place when I was working with bees on my own. I had been long encouraged to write a book to share these experiences, but writing does not come easily to me. My wife, Heather, is a wonderful writer and has taken the stories from my past, which are written in the first-person singular "I," and the experiences we have had keeping bees together, which are written in the first-person plural "we," throughout the majority of the text. This explains why the narrative voice changes from time to time throughout the text from "I" to "we."

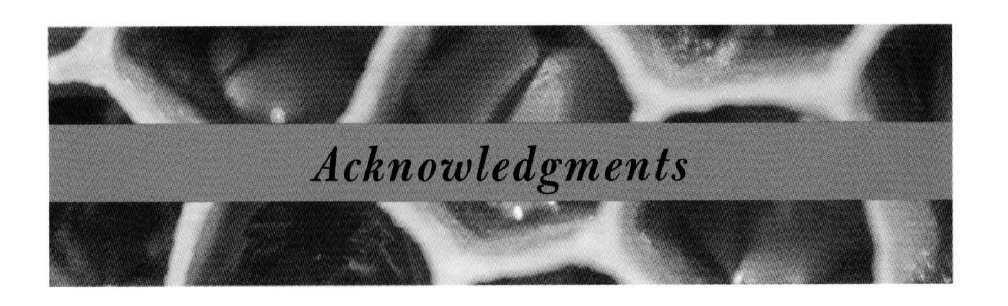

Acknowledgments

It is with tremendous gratitude that we want to thank all of the students, friends, and loving family who helped us to make this book a reality. It was through the support of generous organizations such as the McCune Foundation, the Santa Fe Community Foundation, Prosperity Works, Cuatro Puertas, and the New Mexico Department of Agriculture that we were able to see this project through to its completion. We also thank the Western Sustainable Agriculture Research & Education (SARE) organization for helping us to turn our farm into a demonstration site for pollinator forage species, providing us with the means to give something back to life.

And, most importantly, we must thank the bees. They are our best teachers, and for their willingness to tolerate us as their keepers and share with us their bounty, we are endlessly grateful. This book is for them.

To Them Both

I can remember my young fingers holding a tadpole,
caressing the slime skin with a probing awareness
of life. There was an art
to the preservation, a deep inquiry
through the tunnels of the body
to know as is and not as would be broken.

I remember a wasp on fingertip only with fear
would sting, and the snapping turtles
who stole away toes in the myths
of street children, peering at me with wizened
faces from beneath backs hardened
to the penetration of sun.

I remember a baby bird lost
from its mother's nest, flailing in fallen
leaves, in the hardest conjunction
before independence. We placed it
in a shoebox with a pillowcase and a worm and it died
there before seeing its first dawn of world
or discovering the weight of wings.

I remember ants by the millions
in villages every few feet. They carried
in dutiful progression, the summer into fall.
I watched the tiny pieces of this lifted
parade without the slightest
consideration of what might be
their rightful due in accolade.

But then everything was a prayer,
unconscious and subliminal, like dreams
on liquid space. I'd perch in time, resting
my chin on a palm of salient inquiry, with a child's
full understanding of everything as is. God
was not a word then but genesis was everywhere,
called to me by leaf and brook and breezes
scented by decay.

From these memories I know that life
is truly wondrous and what dies
is what gives birth and to death
I am grateful, as to life
I am living, and one without the other
seems to me our world's very worst
miss of giving.

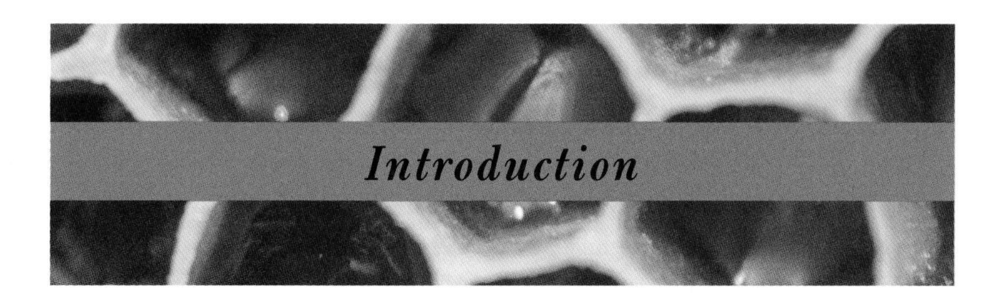

Introduction

For as long as I can remember, I've been fascinated with the world of insects. As a toddler I would watch bees surf from flower to flower, drawn in by the electric buzz of their wings. I meditated on the neon red and blue dragonflies that hovered over our pond, studying their beauty and delicacy. My room had jumping spiders living between the window and the screen, and I'd watch them spring great lengths to grab flies, crushing them with their bionic grip.

By the time I was six, I had my own insect identification book, which I carried with me and leafed through until the pages were dirty and dog-eared. My book showed me that the wasp I thought was stinging our backyard peach tree was actually doing the tree a service by depositing eggs into a peach-tree

borer. Incredibly, it could sense the borer underneath the bark and would insert its long ovipositor through the bark and into the borer's body.

My grandfather was my first mentor and nurtured my interest in insects. He loved working in his organic garden and taught me the agricultural principles that have guided me in my career. His garden was like a great wildlife park for my childhood safaris. I was forever catching some new and unusual creature to study, and he encouraged me to learn what each insect and animal was doing there.

When I was a teenager, my grandfather saw a swarm of honeybees settle on a piñon tree. It just so happened that my mother had an empty beehive she'd never gotten bees for, so my grandfather donned my mother's bee veil and carefully and fearfully ascended a ladder with a handsaw. I was transfixed. Even now I can remember the sharp scent of the piñon needles blending with the delicious sourdough scent of the honeybees. The basketball-sized core of bees purred, while satellite bees orbited around it, some landing, others zinging away into the distant air currents. They did not attack. Grandfather carefully cut the branch, then lowered it into the box, gently placing the top on with pride and relief. I felt a shift in the currents of my life, as though a door had opened to invite me into beekeeping. I put on a bee veil and began spending hours, days, years, and decades in the company of honeybees.

About a year later, my grandfather became very ill. One beautiful spring morning he wanted to go outside and enjoy his iris patch in full bloom, so I accompanied him. I pulled out the coiled hose and he slowly made his way around the rows of blooms while I waited for him to finish. In front of me, there was a patch of mint in full bloom that had taken over an earthworm bed that he and I had built there years before. I thought of how we'd fed the earthworms melon rinds and compost. A black honeybee was busily weaving from flower to flower, sipping the copious mint nectar. At that time, I had five beehives and one of them was full of gray and black honeybees. I knew where that bee had come from.

At that seemingly inconsequential moment I felt another life-changing shift. I knew that Grandfather and I would taste the nectar from our mint once it was made into honey by the bees. The soil, the earthworms, the traces of melon rinds, the mint roots and flowers, the sun, the water, Grandfather, myself, and the bees all felt connected; I knew that we were all part of the vibrant pulse of warm, beautiful life. I saw that just as mint roots and mycorrhizal fungi tied the soil particles together, the bees sewed invisible threads through the air as they flew from flower to pond to tree to hive, threads that fertilized, completed, and enriched our lives. Although I was grieving my grandfather's illness in my heart, the bees made me feel that all was well. Grandfather would pass on, we would miss him, but he would be fine wherever he went, and he would leave a legacy, inspiring me to walk on his gentle organic path. I felt a strong connection to the great mysterious web of life through that beautiful black honeybee.

My commitment to keeping bees as organically and respectfully as possible began in that moment, but it took many years and experiences to realize. Throughout my life my love of insects has steadily grown, and I have come to consider myself an ambassador between the insect world and the world of humans. My work as a beekeeper is an expression of this love, and I try to communicate first and foremost a reverence and respect for the little creatures that we care for.

When I first began my beekeeping career, commercial beekeeping was like the rest of industrial agriculture, rife with exploitation and chemicals. I needed to see it firsthand to understand it and eventually to advocate against it.

Years after my grandfather's death, I went to work for an energetic man who had 4,000 hives and sold literally tons of honey each year. He had been keeping bees for more than forty years and was a wonderful teacher. I absorbed as much as I could, but I also began to see what I felt were abuses to the environment and the bees. Wide-spectrum antibiotics, bee repellents, insecticides in the stored honeycombs—all of these things went against my feelings about how I wanted to keep bees. I eventually left this job and became a seasonal honeybee inspector for the New Mexico Department of Agriculture.

By the time I became a bee inspector, my own beehives were thriving without the use of any of the chemicals commonly used in industrial beekeeping. I had primarily Langstroth hives but already had begun to experiment with a few top-bar hives. All the beekeeping books of the period stated how important it was to use antibiotics regularly, and although I had used them briefly, I did not feel that they were necessary. Many of my peers felt that it was irresponsible to keep bees without regular antibiotic use, and that if they weren't used, diseases would spread. But there were a few voices in the international beekeeping community saying that antibiotics were detrimental and that beekeepers should be raising disease-resistant bees.

In the late 1970s, Dr. Steve Taber[1] and, more recently, Dr. Marla Spivak[2] began raising strains of honeybees that were disease- and paraside-resistant. This appealed to me as a way of working with nature rather than against it. Instead of using poisons to fight bacteria and wax moths, I saw that we could enhance honeybees' natural tendency to become resistant to disease through selective breeding techniques. I began to select and breed intentionally for disease resistance in my own hives. Now, after thirty-five years of keeping bees, I can state confidently that careful beekeepers can keep bees successfully without using antibiotics to kill bacteria, comb fumigants to kill wax moths, miticides to kill mites, or fungicides to kill chalkbrood fungus.

This book is an offering designed to encourage beekeepers everywhere to keep bees naturally. All over the world, bee populations are declining because monoculture, insecticides, and economics all have conspired to make life in a beehive difficult to sustain. As large industrial beekeeping operations continue to disintegrate, however, there is an opportunity to save the bees in small isolated communities. In backyards and on rooftops all over the world, bees are being kept without chemical inputs. These beekeepers provide honey, beeswax, and pollination to their neighborhoods, thus serving the same function as the black bee I saw that day in my mint patch; they sew together their communities, offering sustenance in a sustainable, life-affirming way. We hope that this book will help this valuable beekeeping community to grow.

CHAPTER 1
Top-Bar Hives

Wild bees will live in any hollow space they can find. I have taken honeybee colonies out of trees, walls, attics, floors, roofs, large wooden spools, and even a long-unused outhouse toilet. All over the world, people of many different cultures have created diverse hive styles to house their bees. Dr. Eva Crane, in her book *The Archaeology of Beekeeping,* describes the many wonderful and inventive ways that beekeepers have housed bees throughout time. Each hive demonstrates its own unique set of advantages and disadvantages for the beekeeper.

Although many spaces make fine homes for bees, beekeepers want hives that can be manipulated to serve our purposes. Ideally, these hives have a uniform shape and size, so that we can move combs from one hive to another without a problem, and can transport hives if needed. Beehives commonly are made with these human industrial needs in mind and throughout history relatively little research has been dedicated to the preferences of honeybees.

The Langstroth hive is a fairly complex hive invented in the middle part of the nineteenth century by a man named Lorenzo Langstroth, and it has been the standard design used in the United States since that time. Langstroth hives were lauded for their revolutionary capacity to standardize the beekeeping industry. Prior to Dr. Langstroth's invention, bees were kept in a variety of skeplike and hollow log structures, many of which required the beekeeper to destroy the hive in order to harvest the honey. Langstroth's invention was a major innovation for the world of beekeeping, as it was a hive

FIGURE 1-1.

with movable combs that could be managed easily, thus paving the way to commercial honey production.

With the major industrialization of agriculture that happened in the middle of the twentieth century, beekeeping became part of the treadmill of chemicals and mechanical standardization that large-scale agriculture entailed. Over time, the Langstroth system has become associated with the use of antibiotics, miticides, comb fumigants, bee repellents, and corn-, sugar-, and soy-based feeds. Wooden and wax parts have slowly been replaced with plastic parts, and the cost of starting a commercial operation is largely dependent on the purchase of expensive pieces of equipment. Small, farm-based operations slowly have been replaced by large migratory trucking operations that use large quantities of fossil fuels and rely on a steady supply of monocrop nectar.

The Langstroth hive, however, does offer beekeepers certain advantages over top-bar hives. Some of these are increased honey production, potentially less maintenance throughout the season, and greater ease of loading and transporting hives for commercial pollination. These advantages can be crucial for the commercial beekeeper, to whom economy and efficiency are a priority. For the first fifteen years of my beekeeping career, I worked with Langstroth hives.

Around 1980 I read an article on Kenyan top-bar hives. The article fascinated me, as it pointed to the possibility of making hives at home with a simple design. At the time, there were a number of things about the Langstroth system that I did not like, and I was interested in developing a hive that was more adapted to my goals of organic maintenance and self-sufficiency.

There were no specific dimensions provided in the article on top-bar hives, leaving the size and shape of my new hives up to me. I experimented with various shapes and sizes, and found a design that worked well (see Figure 1-1). In 1995 I sold my hundred Langstroth boxes and began keeping bees exclusively in top-bar hives. Since I have been a teacher of top-bar beekeeping for many years now, my design has circulated throughout New Mexico, allowing the top-bar community here to exchange combs freely with one another.

Pros and Cons

Often I am asked why I made the transition away from Langstroth hives to top-bar hives. It is important to note that much of what we do as beekeepers will be the same no matter what kind of beehive we use, but there are significant differences in management, depending on hive design.

My primary interest as a beekeeper has been to keep bees organically. The agricultural world has become increasingly poisonous, both to the earth and to the people consuming the products. Although organic standards for beekeeping differ from one country to another, they generally recommend that beekeepers use hives made from natural materials that are not treated with any chemicals. They also commonly recommend that bees build their own combs, forage in an unsprayed environment, are not fed sugar syrup or pollen substitutes, and are not treated with chemical inputs of any kind. The top-bar hive easily accommodates organic requirements. It can be built by a beekeeper for about $30 to $40 out of untreated wood, or even made out of wildcrafted materials such as willow or bamboo. I have even made a hive out of adobe mud. The interior of a top-bar hive mimics a hollow log, and the bees can draw out their combs in whatever manner they choose. The top-bar hive is long lasting and 100 percent biodegradable.

One of my initial reasons for transitioning away from Langstroth hives was the need to store combs over winter. Langstroth hives have boxes called *supers* that are placed above the brood box for honey storage. Once the honey harvest is over for the season, supers are stored for the winter in the hope of preserving the combs for the following season. The storage of supers is a problem because wax moths and mice will eat and destroy the combs. In commercial operations, supers are fumigated with paradichlorobenzene to keep these pests away. Paradichlorobenzene, the chemical used in mothballs, is a known carcinogen that is absorbed by beeswax. The combs in the supers are treated with this chemical and then placed back into the beehives in the springtime to be refilled with honey. In a top-bar hive, however, there is no need to store combs through the winter, which eliminates both the problem and its chemical solution.

Langstroth hives are used in conjunction with sheets of wax or plastic called *foundation*. These sheets are imprinted with a designated cell size, one that's designed to maximize honey production. Although foundation used to be made only from beeswax, industrial beekeepers are turning to the plastic foundation as a longer-lasting and therefore more economical choice for their businesses. Unfortunately, this plastic ends up as an industrial waste product when it is no longer useful. The predetermined cell size of foundation also presents a matter of debate within the organic beekeeping community.

In a natural environment, honeybees will draw out various sizes of cells depending on the needs of the hive, often using a variety of cell sizes on one comb. Bees commonly will build larger cells at the top of a comb and smaller cells at the bottom of the comb. The brood cell size also changes slightly over the course of a season, and the resulting physiology of the emerging bees shifts to correlate with this change. Bees born from larger cells are correspondingly larger and have more fat bodies. Studies have shown that substances in the fat bodies affect worker longevity and that these substances interact with brood pheromones.[1] Bees born from smaller cells, in contrast, are smaller and show greater resistance to varroa mite.[2] In top-bar hives, the bees are free to vary their cell sizes to meet the changing demands of the hive, and there is neither the expense nor the waste of plastic foundation.

The management of hygiene and disease was also an integral part of my choice to use top-bar hives. With the Langstroth system, honey is extracted from the comb by centrifugal force, which leaves the beeswax combs intact. This allows the beekeeper to put them back into the hive to be refilled by the bees. For maximum honey production, this extraction process makes a great deal of economic sense. The bees don't have to rebuild the comb, which uses up a lot of the resources of the hive. The beekeeper thus profits from having more honey at a faster rate of harvest. However, the tendency in large commercial operations is to keep putting combs back into the hive after they've become a potential health risk for the bees in an effort to get as much use from them as possible before they are discarded.

Each time the queen bee lays an egg in a cell, the bee larva lines the cell with its cocoon during pupation. When the new adult bee emerges, the cocoon remains inside the cell, is cleaned out, and a new egg is laid. When the next adult bee develops, a second layer of cocoon is left behind, is also cleaned out, and the process continues. As time goes on, the cell size gets smaller and smaller, and the comb gets blacker and blacker due to a buildup of cocoons, propolis, and fecal matter inside the walls of the cells. The bees born inside these darker cells are smaller, and the hygiene of the hive eventually becomes compromised due to a greater level of pathogens. The temptation for commercial Langstroth operations to continue to use old black combs is great, as the cost of replacing combs means the cost of buying more foundation, as well as losing time and money on having the bees build fresh combs.

In top-bar hive management, a great deal of attention and care is given to cycling out old dark combs. They routinely are moved to the back of the brood nest so that they may be filled with honey and culled. The honey is harvested by taking the whole comb off the bar and crushing it, allowing the honey to drain out through a colander. The bees then have to draw out fresh comb to replace the comb that has been harvested. Although many Langstroth hive advocates believe that it is detrimental to require the bees to make more beeswax continually, the bees benefit from the regular conversion of honey to beeswax. It exercises their wax-making glands, which keeps their glandular systems active and healthy.

Beeswax absorbs and holds oil- and fat-soluble toxins, which are unfortunately abundant in our current environment. As these toxins build up inside the hive, it becomes an increasingly toxic environment for the bees to live in. Continuously rotating and culling combs keeps the hive healthy because it removes wax-soluble toxins and cleanses the hive through regeneration.

Back in 2008, I took a job inspecting bees for almond pollination in California. It was during the time that Colony Collapse Disorder (CCD) was hitting the bees hard. I saw countless dead or empty colonies, full of black combs and with multiple strips of miticide hanging in them. Miticidal strips are meant to be administered for a specific amount of time and then removed and disposed of as hazardous waste. Most people don't know how to go about disposing of hazardous waste, so these strips usually just end up in the landfill. Due to a lack of resources and time, beekeepers were neglecting to remove them at the recommended time, and instead were putting strip after strip inside the hive. While I was there, broodcombs tested in California showed levels of miticidal buildup that had become lethal to the bees.

Miticides are just one of the many chemicals that are wax-soluble. Recent research on the causes of CCD show a wide spectrum of chemicals in beeswax, not just miticides.[3] There is a new class of pesticides being used called neonicotinoids. Bees carrying pollen from a plant treated with neonicotinoids can infect the hive with a potent neurotoxin.[4] Although it doesn't kill the bees immediately, their ability to forage normally is impaired. Whenever we see symptoms of CCD in a hive, we immediately remove all of their pollen stores and move the hive to a safer environment in the hope that it can recuperate gradually.

Lastly, the Langstroth system requires some costly equipment. In order to produce honey in any volume, an extractor is a required expense. Extractors are stainless steel machines designed to pull the honey gently out of the combs through centrifugal force. Although extractors save time during honey harvest and preserve the combs, they have their drawbacks as well. Beekeepers who use an extractor tend to put off honey harvest until there is a large batch of honey to extract. The labor required to use and clean the machine simply isn't justified with small batches of honey. In contrast, as a producer of varietal honeys, it's in my best interest to harvest honey after each plant flow, no matter how large, so that I can bottle and sell unique flavors.

The top-bar system certainly has some disadvantages, the main one being that top-bar hives require a high level of skill and confidence from the beginning beekeeper. The possibility for cross-combing is great, and once a

FIGURE 1-2. Top-bar hives do not require an extractor. In the honey-bottling area, the combs are hand-crushed over a tank with a strainer on top. The equipment can be as simple as a colander and a mixing bowl for small batches produced at home.

top-bar hive has become cross-combed, it can be a very daunting task to clean up. Cross-combing happens when the bees draw their combs at odd angles to the top bars, sometimes spanning many bars at once, which causes the comb to drop and break when the hive is opened. The main skill of top-bar hive management involves spacing the combs so that the bees do not have a chance to cross-comb. On heavy honey flows in late summer, when there are a greater number of bees in the hive, it can be difficult to stay on top of management, especially if a beekeeper has a large number of hives. This makes top-bar hives a potentially discouraging choice for commercial beekeeping. It is also difficult to load and stack top-bar hives for large-scale pollination.

Another clear drawback to top-bar hives is that the bees have to build brand-new combs every time honey is harvested, which reduces honey production. The bees consume 9 to 11 pounds (4–5 kg) of honey to produce every 1 pound (0.45 kg) of beeswax, so the construction of new combs detracts from the amount of honey that gets stored in the hive. I experimented with five top-bar hives and five Langstroth hives one year and found that, on average, the top-bar hives produced 20 percent less honey and six times as much beeswax as the Langstroth hives. Many top-bar beekeepers utilize these higher wax yields by making candles, salves, and balms to diversify their businesses. The economic advantage of greater wax yields may well offset the disadvantage of lower honey production, depending on how beekeepers choose to market their wax products.

The natural top-bar movement has appealed mainly to home and hobby operators who have the time and interest in keeping bees intensively. Although there are some commercial top-bar operations across the United States, they are a small minority compared with the number of Langstroth operations. Most beekeeping publications are designed to serve the Langstroth community, and there is a great deal of support for beginners. Although there are support systems for top-bar hive enthusiasts, they tend to be small groups oriented toward lively discussions of hive design and home-built systems for harvest and wax utilization. Thankfully, this is slowly changing, aided by the Internet and a growing body of resources for top-bar hive beekeepers (see the Resources section at the end of this book).

Top-Bar Hive Design

Bees always go to the top of the space they are going to live in and build combs from the top down. In a top-bar hive, the top bars are the roof of the hive and the foundation upon which the bees build their combs. As beekeepers, we need the bees to build one comb centered on each top bar so that we can manipulate the combs easily.

A top bar should be just the right width to hold one broodcomb while maintaining bee-space on either side of the comb. Bee-space is the space between combs that bees use to do their work. If allowed to build combs in a totally undefined space, bees always will build their combs with a uniform amount of bee-space between them. The top-bar hive is designed to maintain this space between each of the combs.

The width that honeybees build their combs is fairly consistent, although there is some variation depending on the race of the bee. For most European and Asian honeybees, the top-bar width of $1\frac{3}{8}$ inches (35 mm) works well to provide the right amount of space. In areas where the Africanized honeybees are being kept, hives with a slightly smaller top-bar width of $1\frac{1}{4}$ inches to $1\frac{5}{16}$ inches (33 mm) is sufficient. In a well-maintained top-bar hive, the combs will be laid out along the centerline of each top bar, and the bees will be relatively undisturbed as each bar is lifted. If the width of the top bars is too thin, the combs will overlap onto more than one bar, and soon the pattern of one comb per bar will be lost. The same is true of top bars that are too wide.

In order to get the bees to hang the comb on the centerline of the top bar, a cleat or ridge can be attached. Dipping the tip of the cleat into melted beeswax can help to entice the bees to begin building their combs along the cleat line. If there isn't any beeswax available, rubbing the cleat with honey also can help to attract the bees to this ridge. When first putting bees from a package or swarm into a totally empty top-bar hive, it is a good idea to prepare the first six or seven top bars with cleats. Once the first six or seven combs have been built, hive management techniques described later in the text can be used to ensure that all of the combs are built straight on the bars.

When attaching walls to a top-bar hive, the angle the walls make with the floor has a considerable effect on how well the hive operates. This angle either encourages the bees to attach their combs to the walls or not. The beekeeper wants little or no attachment to the walls, so that each bar can be lifted easily without breakage or the unnecessary work of cutting combs.

I experimented with several designs, including nearly vertical walls and half-round top-bar hives made out of wicker and plastic barrels. I found that the more vertical the walls were, the more the bees attached their combs to them. The problem with tipping the walls out at a very wide angle was that the combs got too long across the top and the hive became more and more shallow, necessitating longer top bars for significantly less comb. Bees like to curve their combs when there is a great length to cover because the curve creates a stronger comb. A longer top bar tempts the bees to do this.

In the semicircular hives, the bees stopped attaching the combs to the walls at the point at which the wall was about 30 degrees from horizontal, or when the angle between the wall and the floor was 120 degrees. I wanted to know what interior floor-to-wall angle gives a trapezoidal hive the most volume and did the math calculations, only to find that this angle was also 120 degrees. Then I realized that there was a wonderful correlation between this geometry and the hexagonal cells that bees use to construct their combs. Each one of the interior angles of a hexagon is 120 degrees. I concluded that the ideal top-bar hive design, providing the least side attachment and the most volume, would exactly mimic the angles of a half-hexagon. I have made all of my top-bar hives using the bees' sacred geometry of the hexagon ever since (see Figure 1-3).

The next question to solve was how deep to build the hive. I found that large combs would get heavy and break off the top bar easily. This makes a mess and kills bees. If the combs are really small, they attach to the top bars well, but the hive has to be extra-long in order for the colony to meet all of its space requirements for brood and honey. Even then, a long, skinny hive doesn't provide the bees with an adequate area to cluster in to keep their temperature well regulated. As a part of my experimentation process, I used standard-sized untreated lumber and built hives with 8-, 10-, and 12-inch (20-, 25-, and 30-cm) boards. The hive with 8-inch walls needed to be more than 6 feet (2 m) long, and the bees abandoned the hive. It seems they didn't like being in such a long, skinny box. The 10-inch hive worked fairly well, and the 12-inch hive could be made shorter, but its larger combs kept breaking. I adopted the 10-inch hive and have stayed with it. The weight and size of the combs are easy to manipulate and care for.

The length of the hive was then calculated to accommodate broodcombs and honeycombs when a colony is at its maximum population. I initially built very long top-bar hives based on the volume of my Langstroth hives. They proved to be larger than the bees could ever fill. I made some shorter hives, and eventually some that were so short that they encouraged swarming before there was a surplus of honey for harvest. Our standard hive is now 44 inches (1.1 m) long. We find that this length is a good minimum. Hives any shorter often fill up in the summer with brood but lack the room to store excess honey. We do not make the hives 48 inches (1.2 m) long because we need to move beehives in trucks every now and then, and two 44-inch hives

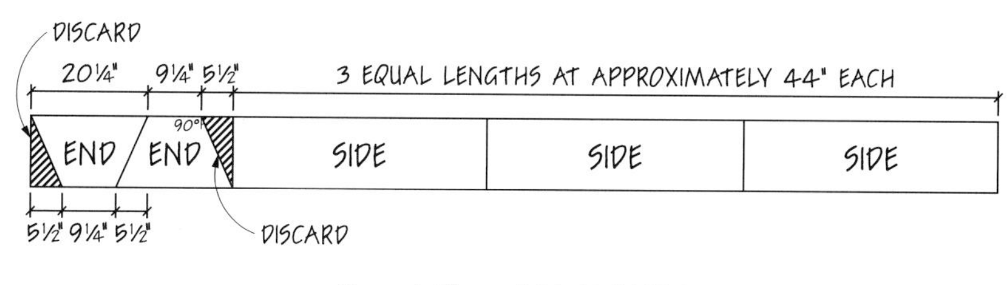

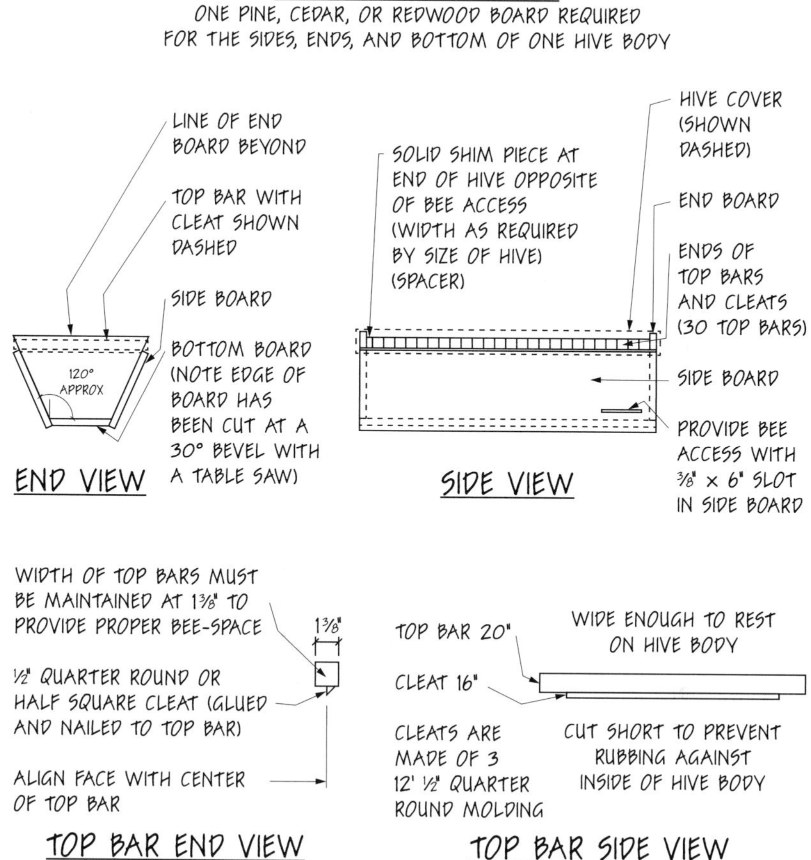

GENERAL NOTES:

1. ATTACH THE COMPONENTS OF THE HIVE BODY WITH EXTERIOR-GRADE WOOD GLUE AND GALVANIZED NAILS OR SCREWS.
2. PAINT EXTERIOR OF HIVE BODY WITH HIGH-QUALITY EXTERIOR-GRADE PAINT. DO NOT PAINT HIVE INTERIOR OR TOP BARS.
3. CONSTRUCT A HIVE COVER WITH MATERIALS AT HAND. CONSIDER A 1 x 2 WOOD FRAME WITH 2" RIGID INSULATION INFILL AND GALVANIZED METAL TOP. SECURE TOP TO PREVENT FROM BLOWING AWAY. BE INVENTIVE!!!

FIGURE 1-3. Top-bar hive design.

fit nicely between the wheel wells. Another drawback to longer hives is that, over time, the boards tend to bow out in the middle.

Another design consideration was the size and position of the hive's entrance. I always had assumed that it should go on the end. But an unrelated experiment caused me to put beehives in a greenhouse. The hives had to be butted up against one another in order to fit, so I put additional entrances on the sides of the hives. When I took the hives out of the greenhouse, I assumed that the bees would resume their use of the entrances at the ends. There were only four hives, but they all continued to use the side entrances and closed off the end entrances with propolis. I tried opening the end entrances several times, and the bees kept closing them. It occurred to me that the location of the entrance could make a significant difference in the way air circulates through the combs. An entrance on the end makes the bees pull air under the first and subsequent combs, while a side entrance allows the bees to pull air in between several combs at once.

I then built some hives with side entrances in multiple locations along the length of the hive. I found that in our New Mexican semidesert climate the bees avoided putting brood in the combs located by the entrance. (It may be that the dry air made it difficult to keep the brood humid.) If the entrance was located at the center of the hive, the bees put the brood nest on one side of the entrance or the other, but there was no predicting on which side the brood nest would be. If the entrance was located at one end, the bees put some pollen and honey in the first few combs and then established the brood nest closer to the middle of the hive. This positioning allows the beekeeper to open the end of the hive away from the entrance, with the assurance that there will be empty top bars or honeycomb rather than the heart of the brood nest. This is important, as it is undesirable to disturb the brood nest any more than necessary, and often when one is working bees, only the very back of the brood nest needs to be observed.

Lastly, there was the question of how big to make the entrance. Langstroth hives have an entrance as big as 16 × ¾ inches (40.6 × 1.9 cm) and an entrance reducer that closes it down to ⅜ × ¾ inch (0.95 × 1.9 cm). Research done by Dr. Elbert Jaycox in 1981 at the University of Illinois indicated that bees that were allowed to choose among many boxes all with the same dimensions and solar orientation showed a preference for boxes with entrances about 4 to 8 inches (10–20 cm) in area that were located near the bottom of the box.[5] Based on my desire to build hives with dimensions that incorporated honeybee preferences, I have chosen entrances of about ⅜ × 6 inches (0.95 × 15 cm). It is just tall enough for a drone to enter, but keeps out predators such as mice.

Many hives now are offered with screened bottoms in order to help with varroa mite control. Studies have shown that if a mite falls off of a honeybee and lacks transport back up into the hive, it will die. Screened bottoms thus keep mites at bay by keeping them from reattaching themselves to bees who are walking on the bottom of the hive. When a hive bottom is made entirely of screen, the bees have a harder time controlling the temperature and humidity within the hive. However, if a screened bottom is placed above a wooden bottom board, the mites will effectively be kept from the bees, and the temperature and the humidity of the hive can be maintained. I do not advocate screened bottoms anymore, as they make for a more expensive and complex hive design at a time when feral bee populations are showing strong resistance to varroa mites. It has been many years since I have been concerned with the presence of varroa mites among my bees, and find that genetic selection is the best defense against this pest.

CHAPTER 2

The Supercreature

A colony of honeybees is a supercreature made up of 3,000 to 80,000 bees. This creature lives by eating and storing the nectar and pollen from flowers; it is in intimate contact with hundreds of millions of flowers in its territory and must carefully use and defend its food because there are times of the year when there are few or no flowers to sustain it. The beeswax combs are like the bones of the creature; they form a framework on which the colony can survive. Inside the combs, food is stored and young are raised.

No single honeybee can live for long on its own. Honeybees are eusocial, meaning that they live cooperatively and need one another to survive. Each bee could be likened to a cell within the human body. Although each bee is necessary and has a unique place within the supercreature, it is also somewhat disposable. Bees, like cells, are continuously dying and being born in an ongoing process of regeneration. Their ultimate purpose is the well-being and continued success of the colony as a whole.

There are three castes of bees in the hive. The queen is the mother of all the bees and thus defines the hive's genetic makeup. The workers gather all the food and care for one other and the young. The drones are responsible for spreading the hive's genetics into the world. Within the worker caste, there are groups of bees that come together to perform specialized tasks for the colony. Each of these groups is somewhat akin to a human organ and is responsible for a unique life-sustaining function within the hive.

FIGURE 2-1. A queen, a drone, and worker bees on capped brood.

Worker Bees

Worker bees do the building, gathering, storing, defending, and decision-making for the hive. They gather nectar to be distilled into honey, pollen to store for protein, water to drink and humidify the hive, and antimicrobial propolis from trees to caulk, disinfect, and varnish the hive. They are all sisters to one another and daughters of their queen. When worker bees first hatch, they perform tasks in the brood nest, an ever-expanding and -contracting area in which all the larvae and eggs are concentrated; often it feels like the heart or source of hope for the hive. These young nurse bees spend their time cleaning cells, feeding the larvae, and attending to the queen. Some workers deliver partially digested food throughout the hive, serving both a digestive and circulatory function. Others serve as a respiratory system, anchoring themselves and buzzing their wings to pull in oxygen-rich air and expel air saturated with carbon dioxide.[1] Still others grow the combs of the hive by gorging on honey and exuding beeswax from glands under their abdomens, while other workers gather the wax from them and form it into hexagonal cells.

As the workers age, their tasks and bodies change. They work further and further from the brood nest, eventually moving to the outside of the hive, where they spend their final days gathering resources from the flowers around them. A worker's lifespan is about six to twelve weeks long in the summer and up to twenty weeks in the winter.

The Queen Bee

The queen serves as the female reproductive organ of the hive. Her sole function is to mate and lay eggs. She is born from a fertilized egg that is no different from the egg of a worker bee, except that it has been given copious quantities of a special larval food called royal jelly. This exclusive diet causes the queen to grow larger than other bees in the hive, so that while she is pupating, a special queen-cell must be built around her. Her extended abdomen houses enlarged ovaries full of eggs.

Unlike worker bees, who will die after giving one sting, a queen has the ability to sting more than once, but she rarely uses this ability. The main purpose of her stinger is to kill her sister queens before they can hatch. When a hive is raising new queens, they often will raise more than one. The first queen to hatch will make a high-pitched piping sound as she surveys the hive for other queens who are still in their cocoons. They will respond to her from within their confines, and she will sting each one to death, thus establishing her matriarchy over the hive.

A few days after a queen hatches, she is led out of the hive by a squadron of matchmaking workers to distant areas where drones congregate in the air. She is escorted to this distant location so that there will be little risk of her breeding with the drones from her own hive, who are her brothers. Inbreeding is devastating to honeybee vigor, so the bees instinctively will travel far to find new and diverse genetics.

Once the queen arrives at a drone congregation area, she mates with numerous drones while in the air. The semen is stored in a bladder called a *spermatheca.* It may take a few flights to fill her spermatheca, but once it is full, her mating days are over. Once she returns to the hive and begins laying eggs, she is not likely to leave again until the hive decides to swarm, and she will be sent back out into the world to start anew.

When she starts laying, the queen will measure each cell that the workers have prepared for her and will then lay fertilized eggs in worker-sized cells and unfertilized eggs in the larger drone-sized cells. Most of the eggs she lays will become workers and will reflect the genetics of the many drones she has mated with. A good queen can lay up to 1,500 eggs a day at the peak of the hive population and will lay hundreds of thousands of eggs throughout her lifetime.

Although the queen provides the hive with all of its life force, she is also entirely dependent on the hive to live. She cannot feed herself and is constantly being attended to by a cadre of workers. She is truly like an organ that if removed from the hive without life support would immediately perish. She doesn't have the glands necessary to digest food the way that workers do, so her meals are predigested for her, then fed to her as she works.

The queen's pheromones have a very powerful effect on the hive. They suppress ovary development in the workers, keeping them from laying eggs of their own. The brood, or baby bees, also emit a strong odor, which keeps the workers performing their functions as nursemaids and providers.

When a hive becomes queenless, it will have a profoundly different energy inside. Often a beekeeper can tell that a hive is queenless simply by the sound of the buzzing. Hives without a queen have a sad, low tone that can be likened only to the sound of mourning. The bees seem to feel hopeless, and without the intervention of the beekeeper the colony will die.

Drones

Drones are the male component of the hive. When flowers are blooming in abundance and the swarm urge is strong, the colony invests in its male genetics by raising a few thousand drones. Drones are haploid, or fatherless. The queen withholds semen when she lays an egg in a drone cell. Drones have larger eyes than workers, and a thicker rectangular body whose abdomen is stingless. They, like the queen, cannot feed themselves and must be cared for by the workers. The sole function of a drone is to spread the hive's genetics into the world, and as soon as drones are ready, they will leave the hive to mate.

When nectar flows cease, the hive withdraws its male aspect by no longer raising drones, and even will evict the drones it gave birth to earlier in the season. Hives that are low in resources cannot afford to feed drones, and will be seen throwing drones out, sometimes even before they have fully pupated. When there is no fresh nectar or pollen coming in, the hive wants to feed only the queen and the few thousand workers it needs to survive.

THE COMPONENTS OF THE HIVE

FIGURE 2-2. Comb that is empty except for the tiny eggs that a queen has recently laid inside each cell. It takes some time to train the eyes to see eggs–an important skill in beekeeping.

FIGURE 2-3. The eggs turn into larvae and are then capped with wax to finish their pupation. This photo shows the various stages of larvae, along with capped brood on the right.

FIGURE 2-4. Worker brood has a flat surface and drone brood is raised slightly in little domes. This photo shows worker brood all around a cluster of drone brood in the bottom center.

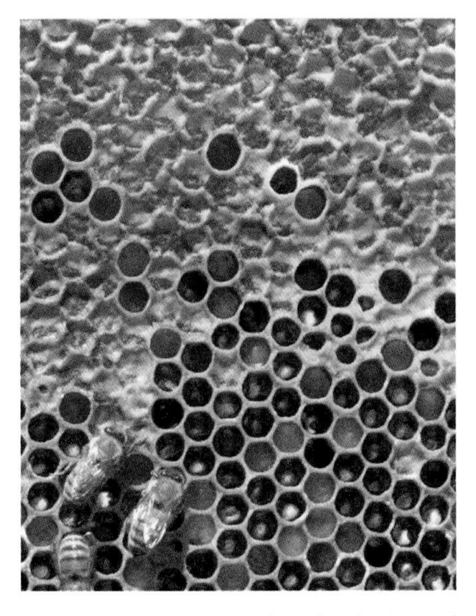

FIGURE 2-5. Capped honey often is found at the top of a comb, and cells of pollen are found dappled below it.

CHAPTER 3
Beekeeping Basics

There are times and conditions in which it is best to open a beehive, and a good beekeeper will respect the boundaries of honeybees by conforming to some simple rules of etiquette. It is best to open a beehive when it is warm, calm, and sunny, and when flowers are blooming. Most often this will be a spring or summer day in the late morning or in the afternoon, when many bees are out conducting the business of the hive. The best conditions for checking a hive occur when there are healthy nectar and pollen flows taking place, and the bees are in a place of abundance. Their protective instinct will be low, and their focus will be on expansion rather than defensive isolation.

In contrast, when it is cold, windy, or about to rain, all the bees will be inside the hive keeping themselves warm and dry. Rather than being preoccupied with gathering nectar and pollen, they will be more interested in preserving the delicate ecosystem within the hive and will resent any disturbances. Often hives that are good-natured on a bright sunny day will show aggressive tendencies if a beekeeper tries to open them in adverse weather conditions. The older foraging bees are the most likely to sting, so it is best to enter the hive when they are away at work, not at home waiting to be mobilized for the common good.

What a beekeeper wears into the beeyard also can affect the behavior of the bees. Although the classic white suit is not necessary, it is helpful to have light-colored, smooth-textured clothing. The less we look and feel like a hungry bear, the better. The presence of dark-colored hair or fuzzy materials will aggravate the sting response. Research has shown that the color red is similarly aggravating to bees, so red clothing also should be avoided. Often we wear t-shirts and jeans to the beeyard rather than protective suits, simply because they are more comfortable. Many of the beekeeping suits available on the market are made of heavy materials, which on a hot day in summer can become almost unbearable to wear. We prefer to use techniques that prevent the sting response rather than putting up physical barriers between ourselves and the bees.

The bee veil, however, is the most important piece of beekeeping equipment, and we advocate always wearing one. It covers the hair on our heads and protects our vulnerable eyes, nose, mouth, and ears. Mammalian breath can aggravate the sting response and provide a target for the bees to move toward. The veil keeps the bees from entering the mouth or nostrils where they are likely to sting. It is quite possible to open a docile hive on a warm, sunny, fragrant day when flowers are in rich bloom and look through the hive without a veil, but it is not recommended. A sting to the eye, nose, or mouth can be very painful and debilitating. Often beekeepers will feel a sense of pride that they no longer wear a veil, but there is no shame in taking respectful precautions when handling honeybees. One strong exhale can bring a guard bee racing toward the mouth, and then it will be too late to put the veil back on.

The Sting

The principal reason that many people never choose to keep bees is due to their fear of bee stings. Many of us have deeply embedded memories of our first sting and the searing hot pain that accompanies it. Despite the pain of a sting, beekeepers have persevered in their work and have learned that honeybees have a predictable behavioral pattern. By trial and error we have found that certain behaviors, timing, scents, or textures intensify or diminish the sting response when we open a beehive and take out the combs.

Honeybees must defend their hive. A wide variety of predators, from insects to bears, would love to eat the contents of the hive. The honey, cultured pollen, and plump bee larvae provide a gourmet meal for many

FIGURE 3-1. A honeybee with her body curled and her stinger extruding in preparation to sting. Illustration by Christopher Clow

creatures. Because honeybees must survive times of dearth, when there is nothing for them to eat, they must vigorously defend the reserves they have stored throughout bloom time, or else they will starve.

Among the tens of thousands of worker bees with stingers tucked away in the tips of their abdomens, there are bees whose primary job is security. The guard bees patrol the entrances to the hive and waft an alarm pheromone into the hive if they feel, see, or smell the approach of an intruder. This olfactory alarm is dispersed quickly throughout the hive. Similar to the presence of adrenaline in human bodies, this scent initiates a fight-or-flight response in the hive. The older bees drop what they are doing and go to the entrances to join the guards in the possible fight. The younger bees begin filling their honey stomachs with honey in case the intruder is too powerful for the defenders and they must flee with the queen to establish a new hive elsewhere.

If an intruder tries to break into the hive, the guard bees initiate a defense response. Often they will begin their defense by bouncing against the face of the intruder as a warning, but if their defense response is fully triggered, they will immediately begin to sting. Honeybees instinctively will attack the eyes, nose, mouth, and ears of an intruder, in an effort to inflict pain where it will hurt the most and interfere with the animal's ability to see, hear, or smell. If all goes well for the bees, the intruder will run away, not too many bees will die, and the hive will survive with the honey it needs.

Once a bee plunges her barbed stinger into the intruder's skin, she cannot pull it out. The whole stinger rips out of the bee's abdomen and the bee bleeds to death. With the stinger, she also has left behind an alarm pheromone, which serves as a targeting device for other bees to aid them in their location of the enemy. A stinging bee can be swatted away, but the firmly implanted stinger is designed to slowly push itself in deeper and inject more pain-inducing acidic venom. The venom sac, attached to the stinger, continues to pump venom while the muscles in the stinger continue to contract, driving the lancelet deeper into the skin. The most effective way to reduce the pain and swelling of a sting is to scrape the stinger away from the skin carefully and quickly, removing both the sac and the barbs in the process.[1] Then the area should be smoked to cloak the alarm pheromone that has been left on the skin.

When a bee is stimulated to sting, it helps to maintain composure and quickly remove the stinger in order to minimize the amount of venom it can inject into the skin. There are many substances that beekeepers advocate to relieve the pain and swelling of a sting, such as honey, baking soda, mud poultices, and commercial antihistamines. Although many people may swell considerably when first stung, often this response diminishes over time. Some beekeepers find that the response, rather than diminishing, becomes stronger and affects parts of the body away from the sting site. This constitutes an allergic response. Although even significant swelling is normal, any kind of immune response away from the sting site is potentially dangerous and can prove to be fatal if it isn't addressed. I have known many beekeepers who have chosen to be desensitized to bee stings, which is a fairly simple medical procedure that involves receiving increasing doses of venom over a period of time until the allergy is gone.

Ultimately, our fear of being stung (if we are not allergic to bee stings) is simply a fear of pain, and this can be overcome easily. The joys of forging a trusting relationship with honeybees is well worth the risk. If we're careful and considerate of our hives, we can offer as much to them as they offer us in exchange, and a long relationship of love and respect can be established between us and our honeybee allies.

Lighting a Smoker

There are cave paintings thousands of years old depicting people wafting smoke through honeycombs. The gentle use of smoke greatly inhibits the sting response in honeybees. By puffing a few puffs at the entrance of the hive, and into the back of the hive once it is opened, a beekeeper suppresses the overall sting response of the bees. The smoke serves to cloak any alarm pheromone, which is released into the air when the hive is disturbed, but it also may draw on the flight response of the bees in the presence of fire. Historically, bees have lived mainly in trees, so the threat of a forest fire would be potentially devastating to their survival. When they smell smoke, they immediately begin to fill up with honey in preparation for a possible escape to a new home. This serves the same function as opening the hive when the older foraging bees are out doing their work; it preoccupies the bees with other activities so that the beekeeper can move through the hive without much detection.

When using smoke, less is often more. Honeybees don't have eyelids, so if their environment becomes too smoky, they will begin to feel irritation and discomfort, and will take to the air. The smoke is used primarily as a signal or warning to the hive, and is not meant to be used to engulf the hive completely. Too much smoke, rather than keeping the hive from stinging, can move the hive into a place of panic, making the beekeeper's job more difficult by creating an uncomfortable environment for everyone involved.

When I first started keeping bees, I had difficulty keeping a smoker lit. I got a job with a commercial beekeeper and watched him light the smoker in such a way that it stayed lit and produced thick, cool smoke for hours. He used dried cow manure. Traditional smoker fuels include burlap; cow, sheep, or horse manure; pine cones; sumac seedheads; coconut husks; or any nontoxic substance that can be induced to smolder gently. If a smoker is shooting flame, it isn't burning appropriately. We want cool smoke, not high-intensity heat.

It is very important to be careful when lighting a smoker. Beekeepers have occasionally caused grass and forest fires. We need to remember that smoke can hide small wisps of flame, and light our smokers away from dry, flammable materials. It is also important to bury or douse any coals completely when emptying a smoker, rather than leaving them on the surface of the ground, where they may reignite in the breeze.

To light the smoker, we first break up enough fuel to fill the smoker. Even when working only a couple of hives, it is better to fill the smoker completely with fuel. Having the smoker full keeps drafts from causing the smoker to go out, and it also cools the smoke by forcing it to move up through the unlit material. Any fuel that is left unused can be emptied out and saved after the work is done. We initially use an igniter fuel, such as dry grass or paper, which we light over the entrance of the empty smoker. Once it is lit, it is pushed into the smoker with the flaming side down and about a quarter of the main smoker fuel is quickly put on top while the bellows of the smoker are puffed vigorously. The bellows must be pumped until the igniter fuel is completely burned up and the smoker fuel has begun to smolder. During this process, the top of the smoker remains open and flames may shoot out of it, so it is best to keep it away from the body and face. It is good to have a lot of flame and heat to make sure that the fuel is lit adequately before more material is placed over it. Once the small amount of fuel is lit, the smoker may be filled, and the bellows may continue to be pumped with the lid closed, until a steady stream of cool smoke is seen wafting from the cone.

Hive Placement

Before getting bees, one should find a good site for the hives.

Honeybees love the sun. Rudolf Steiner, the founder of biodynamic farming, describes them as solar beings. They see pathways in sunlight, and use them to navigate to and from their hives. They respond intimately to the movement of the sun through its seasons, and they feel their greatest life force and abundance when the sun is bringing growth and bloom. So the best location for a hive is in full sun. Hives that are kept in shade may mean that bees spend less time foraging and make less honey. However, in a very hot, sunny climate, where the temperatures regularly get above 100°F (38°C), beehives should be placed in a location with partial or afternoon shade so that they can expend less energy keeping the hive cool.

Honeybees like to have their hive entrance facing south in the northern hemisphere, and north in the southern hemisphere. With this orientation, in a top-bar hive, the length of the hive warms in the morning when the sun hits it. They prefer entrances that do not face into the prevailing wind, and in windy locations they benefit from windbreaks. They also prefer a quiet scene out their door rather than a lot of motion and activity. A screen of vegetation or a wall or fence can be used to block a hive's view of a busy area, providing it is at least 1 yard (1 m) from the hive, giving the bees room to fly up and out.

Beehives should be kept up off the ground. This will prevent problems with predators such as mice, skunks, and ants. Skunks like to provoke the bees to come out at night and will wait at the entrance, eating them as they try to defend themselves. But if a skunk has to reach up to get to a hive, thus exposing its belly, it will leave the hive alone.

Tropical ants will attack and destroy beehives, and common ants will try to rob the hive of its honey. Placing the feet of a hive stand into cans of oil or covering them with a sticky substance can prevent ants from gaining entry. Raising a hive off the ground also prevents termites and soil dampness from decaying the hive. It is best to elevate the hive such that grass cannot grow up over the entrance and snowfall will not bury it. We like to position our hives so that they are at a convenient tabletop height, making them easy to work.

A clean and renewable water source is very important for the bees. They cannot swim and will drown in open containers of water, so it helps to provide them with a container full of rocks in shallow water, so that they can land on the rocks to drink. Ideally, there is a natural source of water nearby, but if there isn't, it is important for the beekeeper to provide water. One of the main nuisances that bees can create for neighbors is monopolizing hummingbird feeders or dog bowls in an effort to get sugar syrup and water.

I once sought out a beeyard for making desert sage, a honey made only in the southwestern part of New Mexico, which is very dry. I drove around an area that looked good and tried to find homeowners who might allow me to place bees on their property during the nectar flow. I found a home that looked promising and drove up to the trailer, where a mean-looking dog was lying in the shade. I managed to make friends with the dog and made it to the front door. I knocked and a man answered, eyeing me suspiciously. I explained that I was looking for a place to put some bees in exchange for some honey.

His first response was, "No way. I already have a bee problem. Just look at that dog's bowl! He can't hardly drink during the day, and it gets hot here." I looked where he was pointing, and sure enough, there were dozens of bees surrounding the bowl, and many floating dead inside it.

"If I can fix that problem, will you let me put bees here?" I asked him.

"Well, sure," he said, looking doubtful. "How you gonna do that?"

"I'll set up a better watering device. You'll see. I'll be back later today."

I left there and went to town to get a tank, a pan, and a stock tank float for feeding cattle. I set the tank near his house and hauled water in buckets to fill it to the top. It held 150 gallons. I then attached a hose to the bottom of the tank and attached the hose to the stock tank valve, which I submerged in the pan. The pan was filled with large rocks for the bees to land on. Before I was even finished, bees were covering the pan, drinking.

I called the man a couple of weeks later and asked how it was going. He said, "Great! You bring all the bees you want! I even bulldozed a space for them!"

The watering device had completely solved his dog-bowl problem, and by the time I arrived with my bees, the 150-gallon tank was two-thirds empty.

Many new beekeepers don't realize how much water a hive can consume, and neglect to provide an adequate water source. In urban environments, this can cause problems, as swimming pools, hot tubs, fountains, birdbaths, and other water sources become clogged with honeybee traffic. Water is an important resource to establish before obtaining bees.

Obtaining Bees

The demand for honeybees is rising continuously, and bees are being imported and exported all over the world. We have found that the demand for honeybees in our area far exceeds the local supply. Unfortunately, the importation of bees from one area to another is not always met with success, as bees raised in one area of the country or world may not be adapted to their new environment. There are also issues with importing not only bees, but also pests and diseases from one area to another. Ultimately, the best solution to these problems is to raise local queens and to work on supporting genetic diversity through selective importation of desirable genetics from trusted sources. When beekeepers are just starting out, there are a number of options for obtaining bees.

Packages

Packages of bees can be purchased in many countries. They are sent to the beekeeper in the mail and are made up of a queen and 3 to 5 pounds (1.4–2.3 kg) of bees. The bees are not related to the queen, and the package does not yet constitute a hive. It is more akin to a swarm, except that the bees have not yet developed any true allegiance to their queen.

When the bees arrive, they should be kept cool and lightly brushed with a little water until they can be put into the hive. Hiving a package can be done at any time of day, but there may be some advantage to doing it in the evening, as the impending darkness keeps the bees from flying and gives them the whole night to settle in. This can be helpful if they have had a long journey and appear agitated or demoralized.

The queen is shipped inside a cage with the bees clustered around her. The cage provides a temporary barrier between the queen and the bees until they have adopted her. Because she is unrelated to them, there is some likelihood that they would kill her if she were released prematurely. Her confinement also makes it easier for the beekeeper to find her, and reduces the chance that the package of bees will abscond (leave the hive) when they are released. Even though the queen is foreign to them, they are attracted to the smell of her pheromones, and they will usually stay in the hive rather than leave as a queenless swarm.

The hive should be ready to receive the bees when the package arrives. Ideally, the box already would have some combs hanging in it from another hive, but if wax combs aren't available, then honey or beeswax can be rubbed inside the hive to make it a more enticing place to live. If the hive has any top bars with cleats, these should be placed at the front of the hive near the entrance, where the package will be poured in.

It is rare for a beekeeper to receive a sting when opening a package because the bees are very disoriented and have nothing to protect, so they are not usually defensive. The package can be opened gently, the queen cage removed and hung in the hive, and the bees then simply poured into the hive. The package will need to be shaken quite vigorously to get all of the bees out. Then it is placed just below the entrance to the hive or in the back of the hive, and any remaining bees will find their way to the cluster by smelling the queen and the other bees.

The queen cage usually is designed to allow the bees to gain access to the queen over a period of time so that, by the time she is released, the bees will be familiar with her smell and consider her their new mother. Typically, there is a candy-filled plug at the top of the queen cage with a cap on it. This cap is removed and the queen cage is hung in the hive. The bees will begin to eat the candy, and when they've finished consuming it she will be released. It is important not to remove any cap or plug that would give the bees immediate access to the queen or let the queen out of her cage. Occasionally queens are contained in cages without candy and must be released manually after a few days.

The queen cage is hung in the hive by passing a thin wire or string through the holes and wrapping it just under a top bar near the entrance of the hive. The candy plug should be hanging to one side. The bees will cluster around the queen and begin to build their combs.

After three days, the hive should be checked to see that the queen is out, and then the queen cage can be removed. If there was no candy plug, then this is the time to let the queen out. It is best to do this inside the hive and, rather than shaking her out or releasing her into the air, the plug is removed and the cage is set in the bottom of the hive so that she can walk out at her leisure. This allows her to make a calm and stately entrance without disrupting the mood of the hive.

Sometimes honeybees will decide to *ball* a queen, or kill her. If they choose not to accept this queen, they may allow her to lay a few eggs, then surround her and sting her to death. Once she is dead, they can raise their own queen from the eggs she laid before she died.

The best time to order packages is in the fall or early spring. In recent years there has been a shortage of bees available due to the increased interest in beekeeping, combined with the declining health of honeybee populations.

There are a variety of races available, and it is best to research which race of bee is best for a particular ecosystem before ordering. It helps to ask around about the reputation of various companies in terms of their queen strength and viability as well as the source of their packaged bees. Companies that are moving toward a less chemical approach may have stronger genetic stock than those that are raising bees on a steady diet of sugar syrup and treating their hives regularly with miticides and antibiotics.

The more awareness the beekeeping community creates about the need and desire for bees that are bred without chemical treatments, the more we will see companies springing up to answer this demand. Calling the beekeeping supply companies and asking about their management techniques can spur this kind of intelligent conversation and get them moving toward more natural management approaches.

Catching Swarms

Another wonderful way to obtain bees is by catching swarms. In a typical four-season climate, springtime is swarm time. The first flush of nectar and pollen brings an expansion in the brood nest, and in good years the bees will build their numbers rapidly and feel the need to propagate themselves by splitting their hives. They will begin to raise a new queen, and once she is ready to hatch, they will send out their old queen with a group of workers to start a new hive. These bees will leave the hive in what is called a swarm and will temporarily rest somewhere until their scouts find a new home for them to move into. During this brief period of waiting, the bees are docile and relatively easy to catch and relocate to a hive.

Contacting pest agencies is a good first step toward finding a swarm. Beekeeping groups in the area may also have a swarm list, or a list of beekeepers that they will call when they are contacted by homeowners about swarms or feral hives. Many exterminators and homeowners don't like to kill honeybees and are happy to have a beekeeper come and remove the bees instead.

Catching swarms can be an adventure, and for those of us who love honeybees it is like finding a great treasure. Swarms can vary in size from 6 inches

FIGURE 3-2. A swarm on a fence post. This is the easiest kind of swarm to catch.

(15 cm) to many feet wide, but most of them are about the size of a basketball (see Figure 3-2). Sometimes swarms will yield excellent queens, and other times they will need requeening because the queen is old and laying poorly. In either case, the bees are a valuable find and provide a beekeeper with bees to keep and a head start in building a beeyard.

Before going on a swarm call, it is helpful to call the homeowner with a few questions about what they are experiencing. Yellow jackets often are confused with honeybees, and some simple questions can save a wasted trip. Yellow jackets eat meat and honeybees do not, so if it isn't possible for the homeowner to make the identification over the phone, they can put out a piece of meat and watch to see if it is attractive to the insects. It is also important to determine whether or not the insects are in a swarm or in an established hive. In the heat of summer, hives may beard from their entrance to keep cool, and often these beards are mistaken for swarms. Removing a swarm and removing a hive are very different enterprises, so it is helpful to have the homeowner provide as many details as possible before packing supplies.

When catching a swarm, it is helpful to have a box, or a small nucleus-sized hive to drop them into. Although we have dropped many swarms into full-sized hives, they are awkward to lift into tight spaces. Placing a comb or two from another hive inside the nuc box can help considerably because it makes the bees feel at home once they are inside. If no combs are available, rubbing three or four top bars with beeswax can help. There also are lures that can be put into nuc boxes as a kind of bait. Honeybees are attracted to the scent of lemongrass, and this essential oil can serve as a lure to help the bees identify the box as a good home. The nuc box should be light enough to carry up and down ladders and small enough to fit into snug spaces, but large enough to house a good-sized cluster of bees. We build them with the same dimensions as a full-sized hive but shorter in length (approximately 16 inches, or 40.6 cm). These nuc boxes come in handy all season long for catching swarms and setting combs into as we work through hives. This nuc box will not constitute the swarm's final home; it is merely a place to help transport the bees before they are moved into their full-sized hive box.

A good swarm hunter will have a truck with a ladder, a number of nuc boxes, a handsaw (for cutting branches out of the way), a smoker, a veil, and a bucket full of smoker fuel. A bee vacuum also can be a handy tool if there is electricity available at the site. Swarms are not aggressive, as they have not yet found a home and don't have a hive to guard, so smoking the bees isn't really necessary. However, it is helpful to wear a veil so that bees don't end up caught in your hair or buzzing around your face. The smoker is used primarily to eliminate the smell where the swarm landed. The bees are attracted by the scent that was deposited there as a homing signal. The smoke helps to eliminate this smell and disorient them so that they are more likely to stay inside the box.

Some swarms are easy to catch, like those that rest on a branch at eye level and can be shaken into a box on the ground. Other swarms located high in trees or tangled in a dense bush are harder to obtain and, unless there is a bee vacuum on hand, they may have to be abandoned because they cannot be shaken out of their location (see Figure 3-3). If it is possible to find and cage the queen and then put her into the box, the other bees will follow her in. Otherwise, it is possible to smoke the location heavily while scooping the bees gently into the box by hand. This sometimes can work in situations in which the bees can't be shaken.

Often when a swarm has been moved successfully into a box and the queen is inside, other bees will come and join her. Closing the box immediately after dropping the swarm, leaving only one or two openings that are just wide enough for bees to get in, helps the bees to feel safe, as though they

FIGURE 3-3. This swarm was very difficult to obtain, and the beekeeper eventually gave up. A bee vacuum would have been helpful in this situation.

have found a home. It is also important to leave the box at the location of the swarm long enough that the bees in the air have a chance to find it.

If the bees have decided to stay in the box, there will be a line of bees at the entrance, fanning the air with pheromone to draw their sisters inside. Fanning bees have a very characteristic stance. They stand with their back legs straight and their abdomens lifted high, exposing a gland on the dorsal side. Their wings buzz quickly, fanning the pheromone into the air around the entrance. Often as one bee starts to fan others join in, forming lines that slowly move into the hive.

Once the swarm settles into the hive, the box can be removed. The best time to remove it is nightfall, when all the bees are inside. If this isn't an option, simply waiting until almost all of the bees are inside is sufficient. This requires heavily smoking the place where the swarm first alighted to reduce fly-back. In cases where the hive must be moved during the daytime, a lightweight sheet can be wrapped around the hive to keep bees from flying out. This also works when transporting a hive in the interior of a car.

Sometimes what looks like a swarm is actually a hive that has taken up residence and begun to build combs. If this is the case, the bees will be much less docile and will be more likely to sting when an effort is made to remove them. The cluster is no longer a swarm and will require a different process for removal.

Hive Removals

For those beekeepers who are really adventurous, there are hive removals. This requires a lot of patience and expertise, not to mention a pickup truck, an extension ladder, and multiple demolition devices. The work is much more labor-intensive, but the rewards can be great. There is a tremendous satisfaction in tearing open the wall of an old shed or barn to find a huge hive waiting to be excavated.

This work also has a number of potential pitfalls, such as working in dangerous locations like on roofs and ladders, often with power tools in hand. It can also be very difficult, especially if the hive is located in a place where the queen can take some workers and retreat deeper into the recesses of a space, far beyond reach. Often the homeowner is relying on the beekeeper to remove the hive completely, and unless the queen is taken, the hive simply will regenerate itself. When a wall or hive cavity is first opened, it is important to take the precaution of plugging any escape routes before removing combs. This initial act can make all the difference between the success or failure of the removal. Spray foam can be useful for plugging holes, as can silicone caulking.

Despite the difficulties, hive removal can be a lot of fun, and if done properly, it can earn the beekeeper some money as well as a hive full of bees. Although removal is traumatic to the hive, the alternate fate is far worse because it is usually poisons delivered via an exterminator. A skilled beekeeper can do this work successfully with little or no damage to the bees.

Once a wall has been opened and the whole hive has been exposed, the first step is to plug any holes around the hive to prevent the queen from escaping. Once this has been done, all of the broodcombs are removed gently. If the broodcombs are larger than the hive box, they must be cut into pieces that can be placed gently inside the box. Ideally, the broodcombs will be handled very carefully so that little or no brood is destroyed. This is especially true of worker brood. Often if a hive has numerous combs of drone brood, these may be disposed of, as the hive may have difficulty supporting them as they adjust to their new home. If there are too many combs to fit into the hive box, both honeycomb and drone comb can be culled.

If broodcomb removal is done slowly and deliberately, the bees will stay clinging to the combs and relatively little panic will ensue. The broodcombs are then propped as upright as possible into the hive box. Each comb is rested gently against the one behind it (angled slightly to leave room between them). Moving slowly and gently will give the bees time to move out of the way as the combs are propped against one another. Placing top bars immediately over the combs as they are placed into the hive will provide the bees with a feeling of protection and relative normalcy, making the removal more peaceful and effective. Immediately they will begin to cluster onto the bars and build comb attachments to secure the broken combs to the bars.

If there is honeycomb in the wall, the bees should be brushed from it, and it should be put into a bucket to be used later as a potential source of feed honey for the hive as it is getting established in its new home. Otherwise, it may be processed and eaten.

If the queen is spotted during the removal, it is best to catch and cage her because her presence in the new box is the best assurance that the removal will be successful. The queen will want to stay with her brood, and if the hive wasn't disturbed too much in the removal process, the queen will likely be one of the many bees clinging to the combs as they are placed into the box. Once both the queen and the brood are in the box, the rest of the honeybees will be drawn in by the strength of their smell.

The bees that are not clinging to the combs can pose a problem, especially if they begin fanning the old location heavily and drawing their sisters out of the box. A lot of smoke helps to prevent this. Bees tend to want to travel upward and cluster at the top of the old hive, so smoking this area, as well as stuffing strong-smelling weeds or cloth in this location, can help to repel them from the old site. Beekeepers who do regular hive removals often carry strong-smelling essential oils, such as peppermint or clove, for this purpose. Bees also orient to visual indicators, so stuffing the old hive space with colored cloth or otherwise changing its visual appearance can facilitate keeping the bees in their new hive.

Once the combs have been removed and the hive box is full and closed securely, it is necessary to rest the hive in a place where the bees can easily find it. Having ropes handy is helpful because many removals require some rigging. If the queen is inside the box, the bees eventually will be drawn inside, but it may take a number of hours or at least until nightfall, when the bees have stopped flying. If a hive was 10 feet (3 m) from the ground, leaving the hive box on the ground usually isn't effective. The entrance of the new hive box must be in close proximity to the location of the old hive, which the bees will be orienting to for the remainder of the day.

In order to venture into the realm of hive removals, one must be very confident with handling bees and also willing to persevere to the very end. If the queen gets away, and the bees don't stay inside the new hive, it is possible to return the next day, and the whole hive may have come back to cluster in their old location, providing another opportunity to complete the removal.

Homeowners will want to be sure that this process won't have to be repeated, and if a beekeeper embarks on this kind of mission, it is important to make sure it is finished well. Hive removals are a community service, and the success of these operations will impact how the beekeeping community is viewed by the general public, so beekeepers shouldn't do this sort of work until they're ready to commit to what can be very hard work.

Trapping

When homeowners don't want to have their houses or property taken apart, then setting a trap for the honeybees is the best alternative. Setting a trap is a fairly easy and straightforward process, as long as there is only one entrance to and from the hive. In situations in which the hive is located in a very permeable environment with multiple entrances, such as a hollow tree or an old house with rotting wood, traps usually are not effective. In order for the trap to work, you must be able to isolate the bees in the hive and secure the entrance completely.

Our method is to make a cone out of ⅛-inch (3.2-mm) metal hardware cloth. The large side of the cone is secured over the entrance to the hive, and the small end of the cone is made to be approximately ¼ inch (6.4 mm) in diameter, such that only one bee at a time can exit. The ends of the exit are frayed, so that many small wires protrude all around it. This prevents bees from landing at the edge of the small hole and crawling back in. If the wires are frayed in a random pattern, in all directions, the bees that are flying in cannot land because the wires hit their wings. Instead, they will collect around the large part of the cone, which is adhered to the wall or tree, trying to access their old entrance. This creates a one-way funnel through which bees can exit, but cannot return to their hive.

We use clear silicone caulking, not only to secure the cone to the wall (along with nails), but also to seal the entrance of the hive after the trap has been removed. It often can be difficult to get the hardware cloth to bind securely to the wall, and bees are experts at finding the least little entrance to make their way back to their old home. Clear silicone caulking is an excellent tool for ensuring that there is no way in or out except the end of the cone. It also leaves a clean-looking site after the removal is done.

Then a nucleus hive is placed just above or beside the cone, with brood-comb inside of it to attract the homeless bees. Once the homeless bees begin to collect inside the nucleus hive on the provisional combs, a comb with eggs can be given to them so that they are able to raise a new queen. Placing a caged queen inside a trap hive is rarely effective because the bees are still too close to their old queen to accept a new one.

The nucleus hive must remain in place for at least two weeks to allow all of the brood to hatch out and leave the old hive in search of food. The queen will never come out and will perish eventually from a lack of workers to feed her. This is one of the unfortunate drawbacks to trapping bees.

The homeowner needs to understand that some honeycomb will remain in the wall and potentially attract future hives, although every effort should be made to seal it after the trap is finished.

Trapping and hive removals are much more successful if done at the beginning of a seasonal nectar flow rather than in late summer, after the bees have had time to build up large populations and store a lot of honey. When hives are removed in the springtime, there are fewer bees and occupied combs to remove, and once the bees are relocated into their new hive, they still will have time to store honey for the winter. Trapping hives late in the honey season takes a lot longer to complete, and the hives may have to be fed or combined with other hives in order to make it through the winter.

Often homeowners will acquiesce to leaving a hive in a wall until spring if they understand that the presence of the bees will greatly diminish during the winter.

Local Nucleus Hives

Honeybees are all about building community, and the best way of obtaining bees is to find someone who is already a beekeeper but who doesn't want to expand and is willing to share some bees during swarm season. Some of the best bees are the ones that already have adapted to the area and whose transport from one beekeeper to another builds a community bond that can be invaluable in the future. As beekeepers, we benefit from forming healthy alliances with one another, sharing information, stories, and our love of these amazing creatures. Before ordering packages or attempting hive removals, look to the beekeeping group in your area as a valuable resource that is waiting to be discovered—one that is rich in bees but that also provides a wealth of community warmth and support.

A nucleus hive, or *starter hive* as it is sometimes called, is just a smaller version of an established hive. When purchasing or providing a nucleus hive, there ought to be a laying queen, a comb of eggs and young larvae, two combs of capped brood, and a comb of honey and pollen. This would be the smallest recommended nucleus hive to begin with. If there isn't a strong nectar flow going on, this size hive will need significant feeding in order to be able to build to the twelve combs it will require during the winter. A better but perhaps more costly nucleus hive would include six or more combs of a similar combination. A larger nucleus hive will have more significant resources moving into the summer and can potentially build up strength and make honey in its first season.

Most packages, swarms, and nucleus hives cannot be expected to make honey in their first season. They have much work to do building the combs of the hive and establishing themselves in the local ecosystem. If it is a particularly good year with strong nectar and pollen flows, it is possible to harvest honey from a new hive, but somewhat rare. In a particularly challenging year, with drought or extreme weather conditions, a new hive may never get off the ground on its own and will require extensive feeding to grow to critical size. New beekeepers have to monitor the progress of the hive continuously in the first season, and adjust their management techniques to reflect what is going on in the environment around them.

Top-bar beekeepers may find themselves having to adapt Langstroth nucleus hives to fit into their top-bar hives. This usually requires cutting the combs to fit. Ideally, when preparing to order these nucleus hives, a request would be made that plastic foundation not be used because it is difficult to modify for a top-bar hive. It is difficult enough to cut comb, crushing brood in the process, without having to negotiate cutting tough plastic in the midst of it all. There are some Langstroth beekeepers who are amenable to receiving a top-bar hive body and establishing bees by putting in a queen and doing a *shaken swarm*. This can be a great alternative to ordering a package of bees or cutting up combs from a traditional Langstroth nucleus hive.

Transferring Bees from a Langstroth Hive to a Top-Bar Hive

There are a couple of options for transferring bees from a Langstroth hive to a top-bar hive. The first is the simplest and one that I used quite often when I still had bees in both styles of hive. It is called *brushing a swarm*. It entails finding the queen, putting her in the top-bar hive, then taking the combs of the Langstroth hive and brushing the attendant bees into the top-bar hive. The young nurse bees will remain in the top-bar hive with the queen, and the older bees will return to their original hive. The older bees in the Langstroth hive will then raise a new queen, and the hive can be the source of another brushed swarm in the future.

If some empty top-bar combs can be provided for the brushed swarm, this helps a great deal, but if combs aren't available, then the brushed swarm is treated similarly to a package and monitored regularly until it has enough straight combs to be maintained using spacing techniques.

Another method works well if the goal is to get rid of the Langstroth hive completely. Cut an ample hole (4 × 4 inches [10 × 10 cm]) in the bottom center of a top-bar hive. Take the top off the Langstroth hive and center the top-bar hive over it, so that the hole into the top-bar hive is situated over the combs in the most active section on the Langstroth hive. Completely seal the top of the Langstroth hive so that the bees can move only upward into the hole provided for them.

Again, if some top-bar combs are available, place them over the hole in the top-bar hive to encourage the bees to move up and begin building new combs in the top-bar hive. The top-bar hive also should be completely closed on top.

The resulting hive is a Langstroth hive with a top-bar *super*, or a hybrid hive. The hive can be maintained like this for as long as the beekeeper would like. If the goal is to eliminate the Langstroth hive, the top-bar hive can be removed once there are six to eight combs built on the top bars (so the queen will have ample comb to lay eggs in) and all of the brood has hatched out of the Langstroth hive. During the period that the brood is hatching out of the Langstroth hive, the queen will have to be restricted to the top-bar hive with a queen excluder. Once the brood has hatched completely from the Langstroth hive, it can be disassembled and any remaining honey or wax can be harvested from it. A plug can be fashioned to fill the hole in the bottom of the now independent top-bar hive, or the combs and bees can be transferred to another top-bar hive and this modified top-bar hive can be kept for future transfers.

Moving Bees

There are times when a beekeeper has to move a beehive from one place to another. Perhaps there is a particularly delicious bloom somewhere, or they are bothering a neighbor, or someone tells you they "have to" spray insecticides. Or perhaps someone has hired the bees to pollinate a crop, or the bees have been purchased recently and need to be taken to their new home.

The complexity of the move depends on how far the bees need to be moved. The beehive has bees that are out foraging all day long, and those bees have learned the exact location of their hive in the landscape of the neighborhood. Bees navigate by vectors in sunlight, and will follow the same path home each day. If the hive is moved while the bees are out foraging, they will come home to the same location, only to find that their home is gone.

If the hive is moved only 6 feet (2 m) or so at a time, then the bees will fly around in a confused manner for a little while, but then will find the hive eventually using their sense of smell. An effort should be made to keep the same hive orientation to the sun during this sort of short-distance move. If a turn of the hive is required, it is best to do this gradually, a little bit each day. One method of moving a hive a short distance is to put it on a wagon and pull it a few feet each day.

When a hive is being moved less than 3 miles (4.8 km) from its original location, the bees may return to their old site through a process of patterned recognition and become homeless. When a hive needs to be moved, the beekeeper either must move it a little bit each day, or move it more than 3 miles (4.8 km) away in one night. When a hive has been moved a long distance, the field bees recognize that they are in a new location because they no longer have any familiar visual indicators, so they orient to the new location rather than trying to find their old site.

If the hive needs to be moved in a cart or vehicle, it must be loaded and moved when all the bees are in the hive. Moving the hive in the middle of the day, when bees are flying, will leave thousands of field bees homeless. They will end up clustered in their old location and may become a nuisance. The hive that has been moved also will have lost its valuable field force, slowing its growth and development. Beehives should be loaded into vehicles when it is dark, or so cold that the bees cannot fly. If it is warm, they must be smoked.

Generally speaking, honeybees do not like to be moved. This is probably a good time to wear a sealed suit and gloves. Often bees will wander outside the hive to investigate the commotion and fall to the ground. Boots and pants secured at the ankle will prevent the bees from crawling up and stinging. Honeybees do not see well at night, so rather than wearing typical beekeeping whites, it is better to wear dark-colored clothing. If using light to navigate in the beeyard, it is best to use a red light. The bees will fly to a light source, making headlamps and even flashlights a potential cause of stings.

If there is a large beard on the outside of the hive, moving it may be both difficult and ill advised. The clump of bees will crawl all over the hive and onto anyone touching it, and often clumps of bees will fall to the ground and will be left behind and trampled. If a full, heavy hive must be moved, it should be divided, even temporarily, into two hives for the move. They will be lighter to carry and fewer bees will be left behind.

A hive breathes a lot of air and should not be plugged up with rags or even screen wire mesh because the bees may crowd around the blocked entrance and suffocate the colony within. I have seen people move a beehive a long distance in a car by removing four top bars at the back of the hive and stapling heavy screen wire mesh over the large hole so the bees had lots of air circulation.

The top bars that do not have comb on them can come loose and slip out of place when a hive is being carried or is vibrating around in a moving vehicle. This is one of the best reasons to purchase or build top-bar hives with lids that have sides that extend down over the top bars, securing them in place.

It is best to transport beehives in an open truck so that the bees are not in the vehicle with the driver. But if a closed car is the only vehicle available for the move, a sheet can be wrapped lightly around the hive to keep the bees from escaping into the car. In hot weather, the sheet should first be dampened with water to provide cooling moisture during the move.

If moving bees in an open truck bed during the daytime, netting the bees will be necessary. Shade cloth can be used for this purpose because it has small enough perforations to supply adequate ventilation, while also keeping the bees from flying out at every stoplight and gas station.

Moving beehives in the winter is easier than in the summer because the cold keeps the bees inside the hive. When moving bees in the heat of summer, there is a danger of having them overheat and perish. Topping the hives with ice that slowly melts during the drive can help to keep them cool.

One of the disadvantages of top-bar hives is that they are not designed for transport in the way that Langstroth hives are. They are not as easily stacked, and the combs do not have a frame around them to strengthen them during travel. This makes the combs more susceptible to breakage if a hard bump is hit on the road.

Besides driving slowly and carefully, there are a few other things that can be done to reduce the risk of breaking combs during a move. Bees often attach little spurs of wax between the comb and the side of the hive. In transport these spurs are advantageous because they reinforce the comb, keeping it from swinging back and forth as the vehicle stops and starts. Lifting the combs out breaks these spurs of beeswax. Therefore, it is best to leave the combs untouched in a beehive for a few days before moving the hive.

Stopping the moving vehicle puts the most stress on the combs. If the hive can be loaded with the long axis of the top bars pointing in the direction of travel, the combs resist breaking much better. In this way we have moved hundreds of beehives thousands of miles on fast-moving freeways with potholes and railroad tracks and arrived at our destination with few or no broken combs.

The most complicated move is one in which you need to move the bees more than a few yards but less than 3 miles (4.8 km) and you cannot move them a few yards a day because of an obstacle (a street, river, pasture with cows) or the bees need to be moved within fewer days than the 6 feet (2 m) at a time will accommodate. In this case the hive must be temporarily moved 3 miles (4.8 km) or more away. They should remain at the interim location at least two weeks so that the hive completely forgets the old location. In two weeks' time, most of the field force will have perished and a new generation will have taken over. Then the hive can be brought back to the new location nearer to the original site.

CHAPTER 4

Hive Management

There are many levels of hive management. Top-bar hives are a simple investment, and it is quite possible to build one, get bees to inhabit it, and never look inside it again. By doing this, one has embarked on the most elemental form of beekeeping by simply providing a natural home akin to the inside of a log. From a permaculture perspective, very little effort has been put into the beehive, yet a long-term benefit is available from the pollination services the bees provide. The bees will follow their own natural rhythms and potentially can live for many decades without human interference. If the conditions are favorable, wax moths live symbiotically within the hive, culling old dark combs, and the hive will wax and wane continuously. Of course, it is a rare beekeeper who wants bees only for pollination or simply wants to provide them with a home. Most of us really would like to taste some of the liquid gold that they have to offer.

If there is a desire to harvest honey, then another level of management is called for, and it involves keeping the hive orderly and preventing cross-combing. Bees naturally will orient their combs in a slightly curved shape, and often will draw them out in patterns that don't match the top bars that have been provided for them. If one is starting with an empty top-bar hive and a package of bees, there is a great deal of management to be done in the first weeks after the bees are hived to make sure that their combs are drawn out along each top bar. The hive will have to be checked every couple of days to track the new light-colored combs that are being built. When they are still only an inch or two in length, they can be removed and reattached so that they are drawn out along the center of the top bar (see Figure 4-1). Cleats also can help a great deal with this. (See the Top-Bar Hive Design section in Chapter 1.) Rookie top-bar beekeepers may find it difficult to remove combs and reattach them for fear of doing it wrong. These small combs are very fresh and soft and, when they are handled, the top of the comb will be crushed in order to press the wax firmly onto the top bar. This is okay, and the bees will repair the comb quickly and make stronger and more fortified attachments. It is better to do this kind of reconstruction work before the combs have grown any larger. This is an absolutely crucial step in learning to manage a hive. Once a top-bar hive has become cross-combed, it becomes a nuisance to both the beekeeper and the bees.

Once there are two fully drawn-out combs, it is much easier to manage the hive as it continues to build. This is done by taking partially built combs and inserting them one at a time between fully built combs. This will give the bees no choice but to draw them out parallel to the combs they are sandwiched between. This management practice eventually becomes like second nature, and every time an expanding hive is opened, the last two combs will be reversed. This moves the partially built comb, which usually is located at the back of the hive where the bees are expanding, in between two fully built combs. If there are two or three partially built combs in the back of the hive, they each can be moved into the body of the hive where there are more fully built combs to sandwich them between.

In times of heavy nectar and pollen flows, the bees will build very quickly, and it becomes extremely important to space them by placing one empty bar between two fully built combs, such that the back of the hive might look like this: one fully built comb, one empty bar, one fully built comb, one empty bar, one fully built comb, one empty bar. (See the following illustrations.) If you check the hives once every two weeks or so, it rarely will be necessary to insert more than two or three empty bars into the body of the hive.

Spacing—or sandwiching empty bars or partially built combs in between fully built combs—is the most important part of top-bar hive management. It most often is done at the back of the hive, but during swarm season it also can be done within the brood nest. This is done to encourage the *stretching* of the hive and to give the bees a feeling of space so that they are not as eager to swarm.

FIGURE 4-1. The beginnings of cross-combing at the back of a hive. The two little combs will be removed and reattached to the bar with the correct alignment, then the bar will be placed between two fully built combs in the body of the hive.

When putting space bars into the brood nest, less is more. Just one or perhaps a couple of partially built combs from the back of the hive will suffice. The bees need to keep the brood warm, and too much space is very disruptive to them.

It is important to know when to space the hive and when not to. If there is a strong nectar and pollen flow, then spacing is a vital management tool, but if there is a dearth of nectar or cold weather at night, the empty bars will provide a challenge to the bees. They will lack the resources needed to build new comb and will not be able to cluster comfortably together; also, the queen will not lay her eggs in a place in which they cannot be kept warm. Honeybees always draw out their combs with just enough space to move easily between them, but no farther apart than that. Their hive ecosystem depends on this unique architecture to thrive. Top bar beekeepers need to recognize this delicate balance and use spacing wisely because it forces the bees into closing empty space. Producing beeswax takes enormous resources for the bees, so the amount of spacing should directly reflect the amount of nectar and pollen flow currently taking place around the hive.

Occasionally, rather than building new comb for honey storage, the bees will fatten their combs, often at the top sides of the bars. To manage this, the bees are gently brushed from the area so that the comb can be trimmed and flattened. Then a hive tool is used to *butter* the edges of the comb by using the flat of the tool to cap the comb at the thickness that is desirable for easy management of the hive.

With these techniques, a hive can be maintained so that harvesting honey is fairly easy and painless for both the bees and the beekeeper. A cross-combed hive is an intimidating prospect for the beekeeper because it ultimately involves lifting bars and having the attached combs break and drop. Bees drown or are crushed in the process, and the result is totally miserable for everyone involved. Humane and civil top-bar beekeeping for honey involves the prevention of cross-combing. There is simply no alternative if beekeepers care about the well-being of their honeybees and want to avoid an anxiety-causing mess.

The following illustrations provide a guide for using spacing techniques and creating various configurations of comb as a part of ongoing top-bar management. The top bars are numbered so that it is clear where each comb was moved during management. Notice that there is a thin bar at each end of the hive. These are used to prevent the end combs from being attached to the walls of the hive. These end bars can be made of scraps left over from building hives or simply can be thin top bars that are turned on their sides. It is helpful to have a variety of bar widths available when visiting the beeyard because changes in moisture and temperature within the hive can cause the body to expand and contract, changing the size of the gap at the back of the hive and requiring a different-sized end bar.

Each of these illustrations is designed to reflect a hive whose entrance is on the left side of the hive. We refer to the front of the hive as the left side of the graphic and the back of the hive as the right side where it is opened for management. When managing hives, we always open the hive at the back, away from the entrance, and rarely do we penetrate or rearrange the brood nest beyond the last comb or two. Leaving the brood nest largely undisturbed is better for the overall well-being of the hive. If we see that there is some kind of problem with the brood pattern, we will search further into the nest, perhaps to locate the queen or discover whether there might be illness, laying workers, or some other dysfunction that we need to address. We also will go through the whole nest when making divides or raising queens.

These illustrations are made to reflect a variety of management techniques, beginning with activities that need to be done when first coming out of winter as the nest begins to expand. Many of these management activities will be explained in greater detail later in the text. The illustrations are presented here as a group for easy reference in the beeyard.

SPRING HIVE MAINTENANCE 1

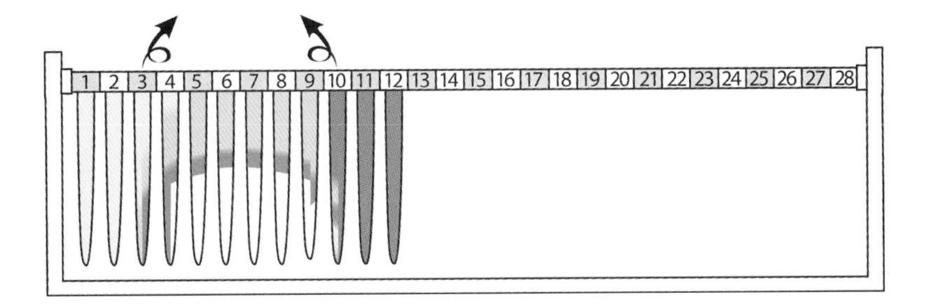

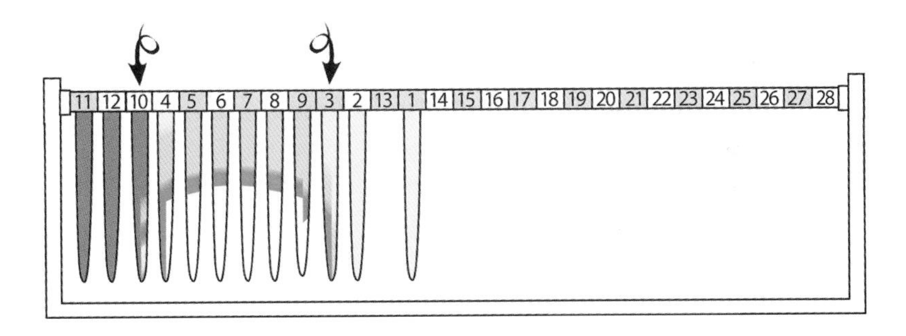

FIGURE 4-2. This illustration represents a hive coming out of winter. The combs filled with crystallized honey create a barrier for the expansion of the hive, so they are moved to the entrance, where they can continue to be utilized in dearth periods but no longer present a barrier to the hive. Note that the empty combs are now at the back of the hive, where they can be filled with nectar, and only one empty space bar has been provided at 13. When a hive first emerges from winter, growth can be slow and nights are still cool, so spacing is used conservatively.

Brood Nest
Ripe Honey
Unripe Honey
Crystallized Honey
Pollen
Empty Comb

SPRING HIVE MAINTENANCE 2

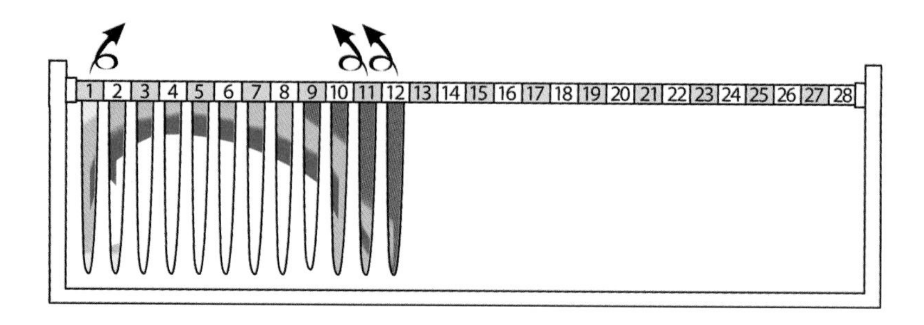

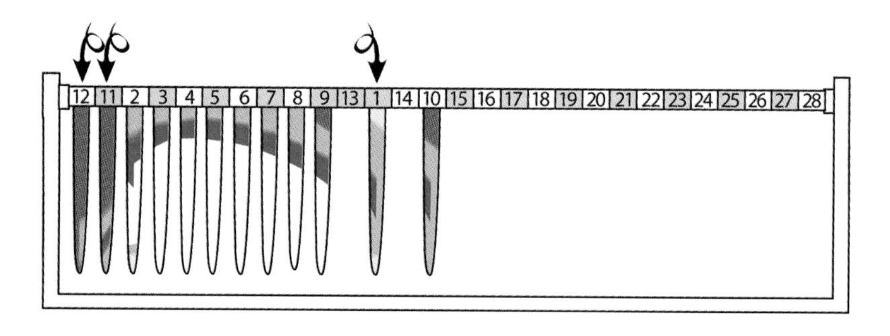

FIGURE 4-3. In this illustration, the hive has come out of winter into a good nectar flow, and there are no empty bars. The crystallized honey is moved to the entrance, and the combs of unripe honey are moved toward the back of the hive. Note that more space bars are provided for the bees than in the last scenario. Because there are no empty combs and nectar and pollen are actively being brought into the hive, the bees will fill these empty bars more rapidly, so more space is provided for them to fill. This kind of spacing also provides swarm suppression by encouraging the expansion of the hive.

- Uncapped Brood
- Ripe Honey
- Unripe Honey
- Crystallized Honey
- Pollen
- Empty Comb

HONEY FLOW

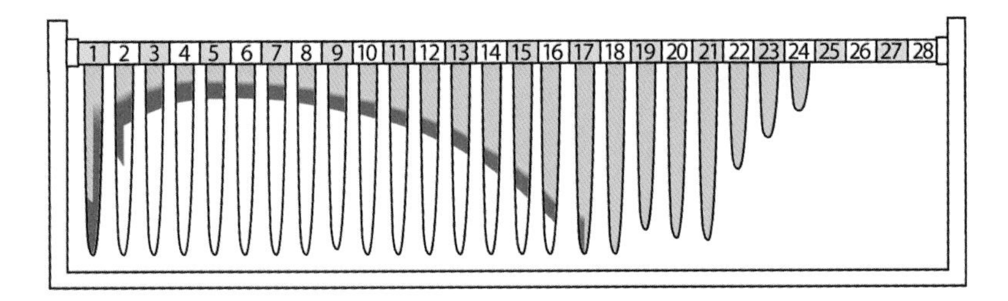

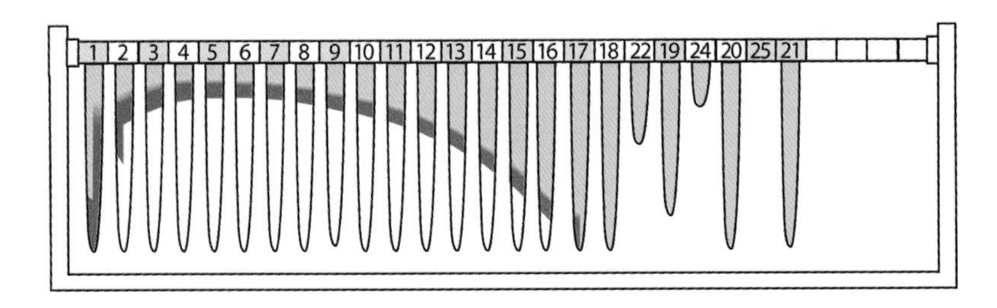

FIGURE 4-4. In this illustration, a good honey flow is under way, and the hive needs considerable spacing. All of the honey is fresh and unripe, and the hive must be managed more closely to prevent cross-combing and swarm tendency. Note how the partial combs at 22 and 24 have been placed between more fully built combs to ensure that they will not be curved as they are completed. This not only prevents cross-combing but also provides swarm suppression, as the worker bees will spend time actively filling the gaps provided for them and will be less inclined to swarm due to space restrictions.

STRONG HONEY FLOW

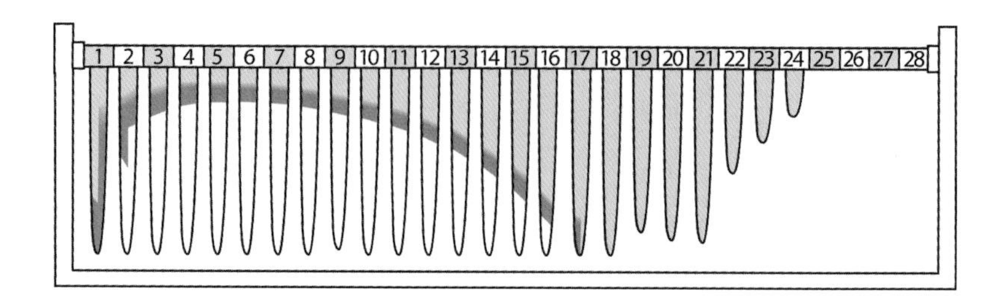

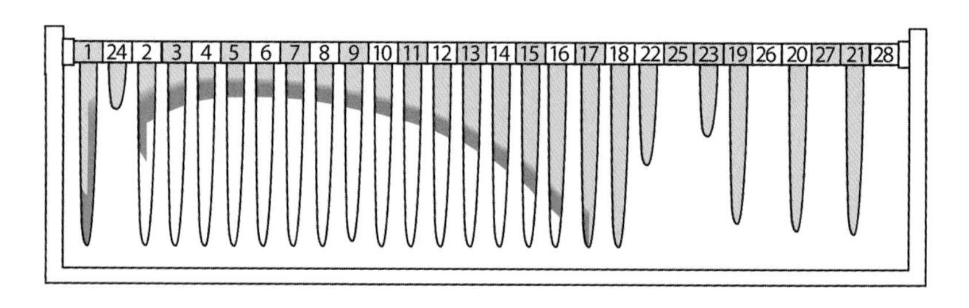

FIGURE 4-5. In this illustration, the hive is on a tremendous honey flow and is filled with unripe honey. The hive is spaced more dramatically, with a greater number of empty bars at the back of the nest and with a spacer bar placed near the front of the hive (24). This kind of spacing provides swarm suppression and encourages the bees to continue building fresh comb. This kind of spacing is used when there is a strong nectar flow coupled with the presence of a good quantity of drone brood, which is the first sign of swarming.

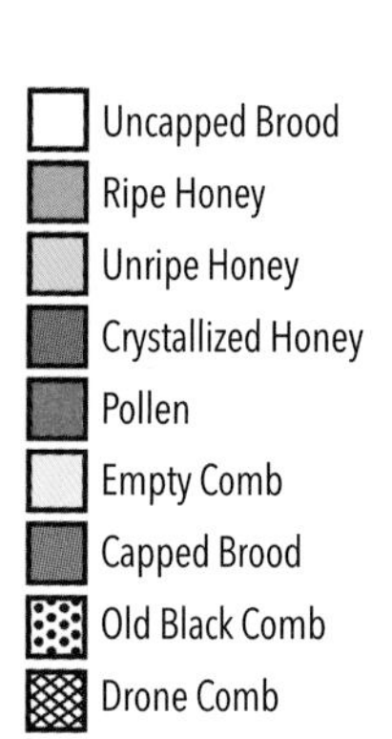

COMBINING HIVES

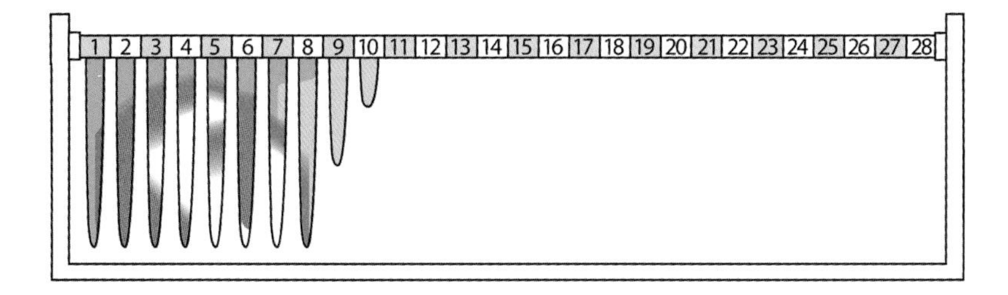

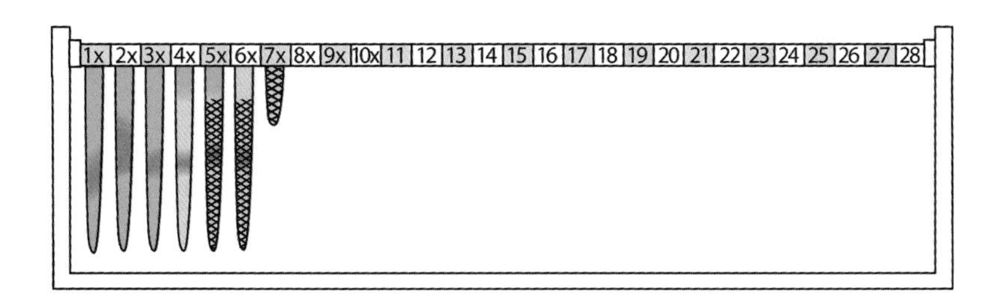

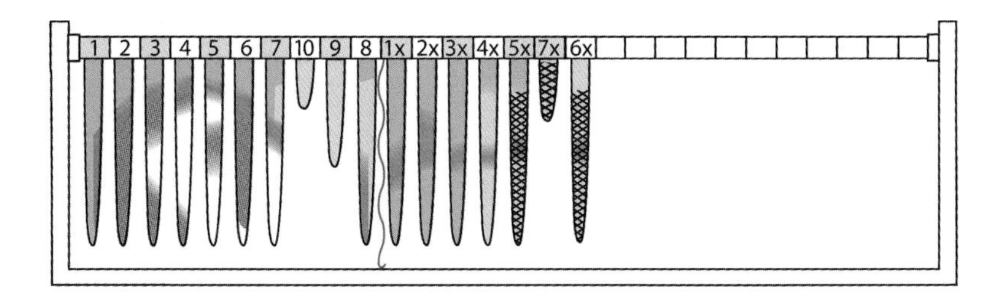

FIGURE 4-6. In this illustration, two hives are combined into one, using a paper barrier, which allows them to gain access to one another slowly, thus preventing a war in which many lives might be lost. The first hive at the top has a good queen but has not built to critical size for winter. The second hive is queenless and has developed laying workers. When combining hives, the weaker hive always is put to the back of the stronger hive. This will give the strength of added numbers to the queen-right hive. The drone comb is placed at the back, where it can hatch out and be filled with honey at a later time.

REVERSE BROOD NEST

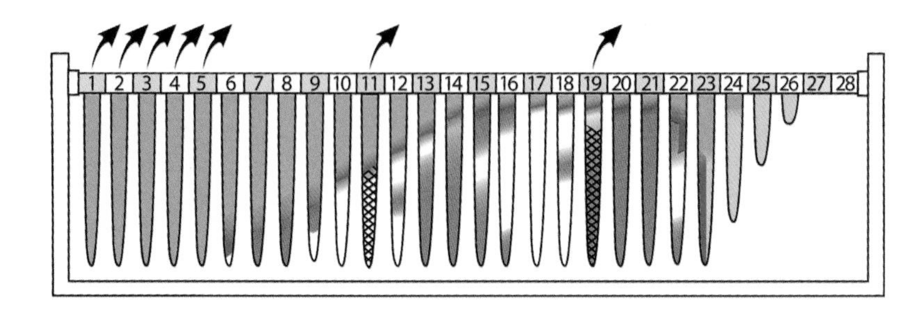

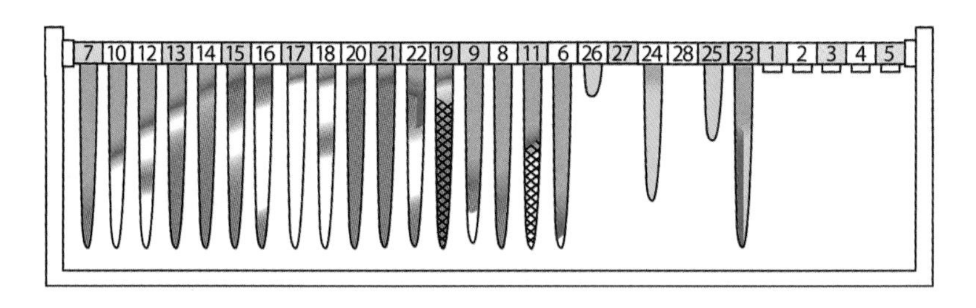

FIGURE 4-7. In this illustration, the brood nest is located away from the entrance, making it necessary to harvest the honey at the front rather than at the back. This is relatively unusual but does happen from time to time. If the hive needs harvesting but there are broodcombs at the back, it becomes necessary to check the front of the hive for harvestable honey. As a part of harvesting and hive management, the drone comb is placed to the back of the brood nest and the hive is spaced for a continuing honey flow. Note how one ripe honeycomb is left at the entrance of the hive (7) and how the other combs that are mostly ripe (8,9) are put toward the back for the next harvest when the brood has hatched out.

HIVE REMOVAL

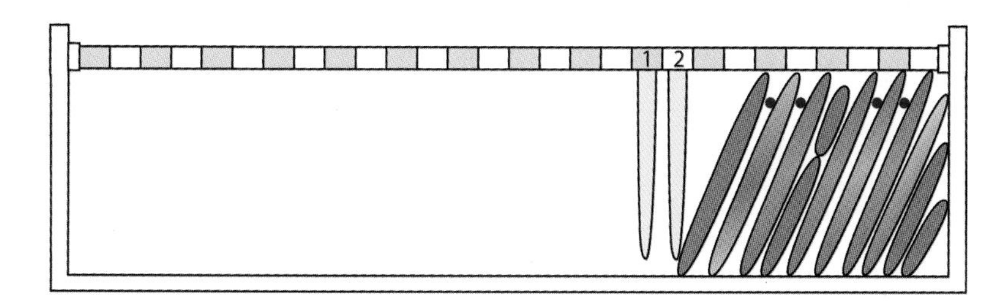

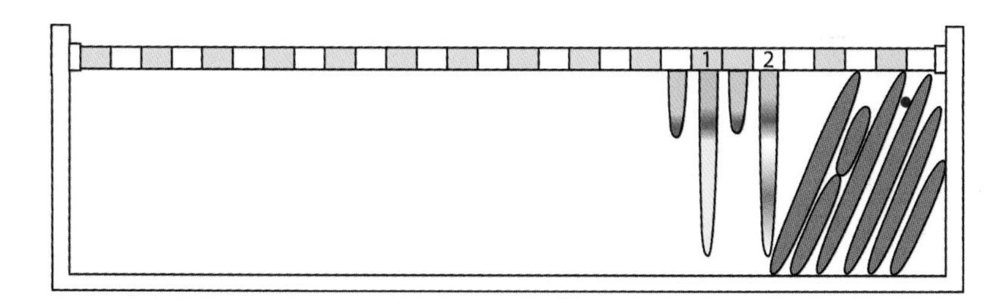

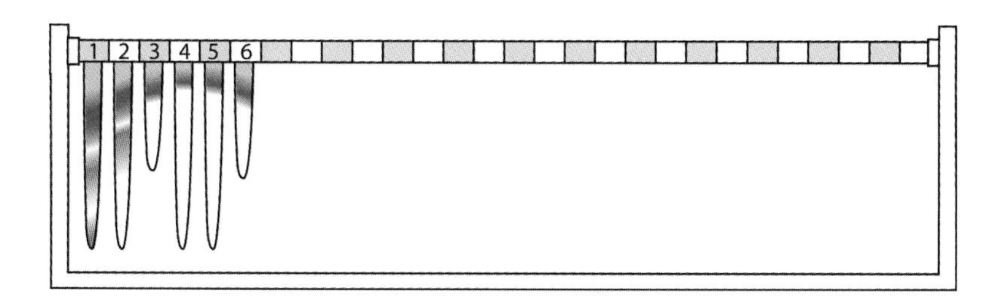

FIGURE 4-8. This illustration represents the evolution of a colony once it has been removed from an unwanted location and put into a hive. The capped broodcombs are propped gently into the back of the hive. (Blue dots represent sticks used to keep them apart.) Two empty combs are hung beside the downed combs. The second graphic shows the growth of the hive onto the bars and spacing techniques to encourage this growth, along with the removal of some of the old broodcombs from which the larvae had hatched out. The final graphic shows the hive with all of the old comb removed and its new location at the entrance of the hive.

BROOD NEST MAINTENANCE

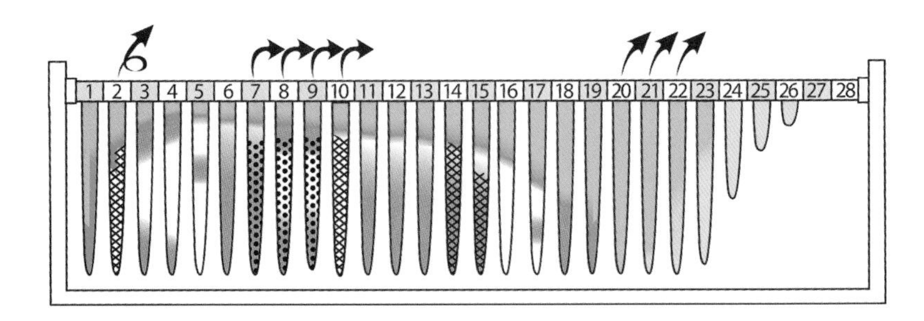

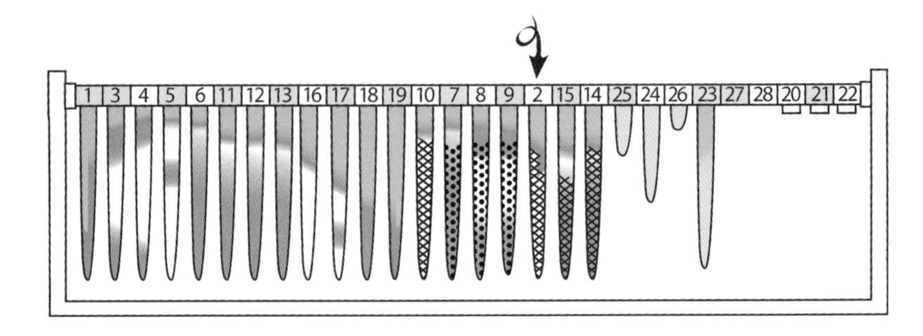

FIGURE 4-9. This illustration shows how to modify a hive to prepare it for winter or a period of dearth and contraction. The old black comb and drone comb are moved out of the brood nest toward the back of the hive, where they can hatch out and be culled. The lighter-colored broodcombs are consolidated at the front to provide a clean nest for the hive to contract into during its period of dormancy. A comb of ripe honey always is left at the door, and all the honey in the first twelve combs will be left for the hive's sustenance during the time of dearth.

QUEEN CELL-BUILDING HIVE

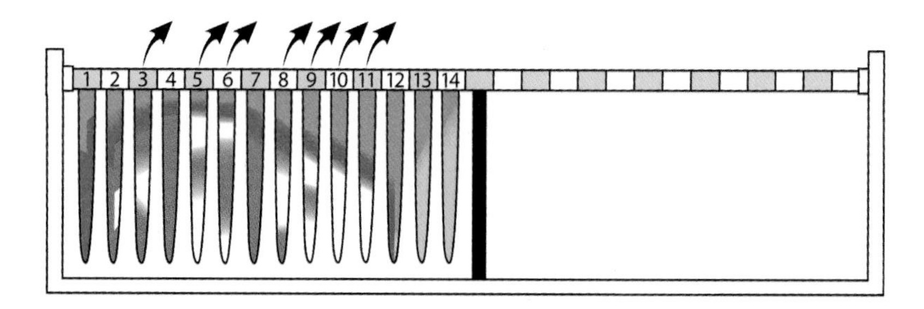

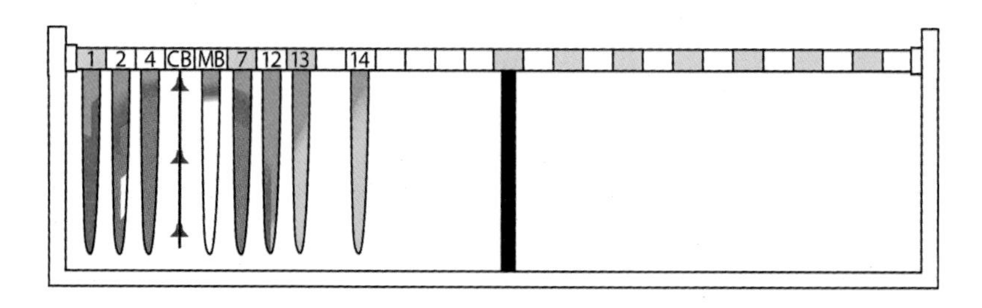

FIGURE 4-10. This illustration shows how a full hive is divided in half with a partition to make a mating nucleus hive. Then the mating nucleus hive is modified to become a cell-building hive for queen-rearing. Much of the brood is removed and given to other hives, along with much of the honey, and the hive is spaced for growth. The queen-cell bar is hung in the center of the brood nest, where the bees will concentrate on its development into a healthy display of queen cells.

TWO-QUEEN SYSTEM

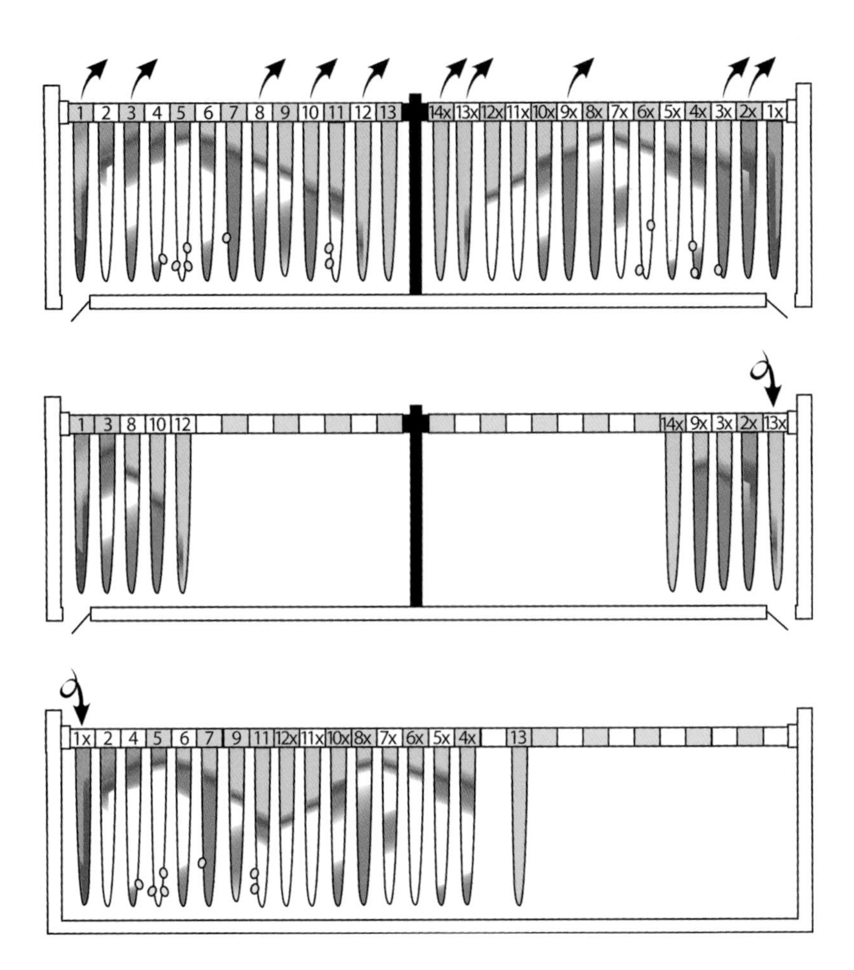

FIGURE 4-11. This illustration shows how to use a two-queen system to make divides and honey in the spring. The double hive at the top has two healthy hives separated by a divider. They have begun to ready themselves for swarming by producing numerous queen cells. Two queen-right divides are made with five combs each of capped brood and honey, and the partition is removed from the original hive, combining the remaining combs and bees. This new combined hive will raise its own queen from the queen cells and rapidly become a strong honey producer. (Queen cells should not be included in the queen-right divides. If there are cells located on combs being moved into the queen-right divides, they can be cut out and used elsewhere.)

MAKING A DIVIDE

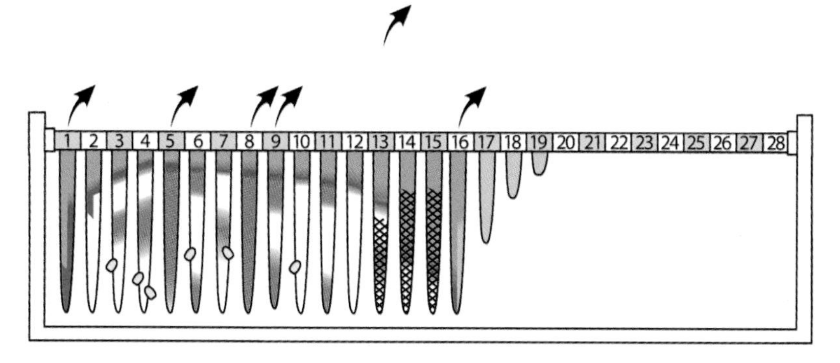

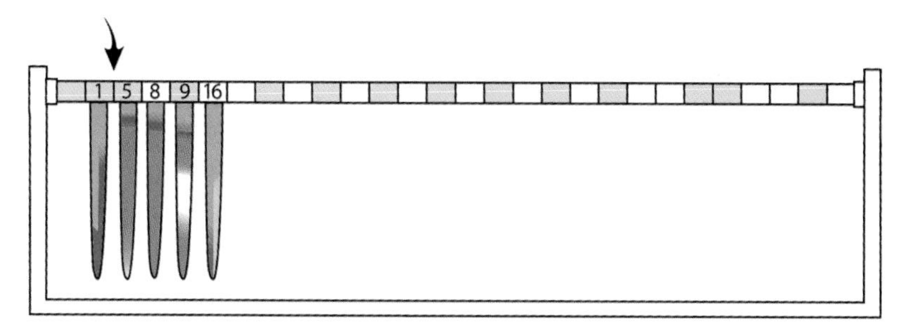

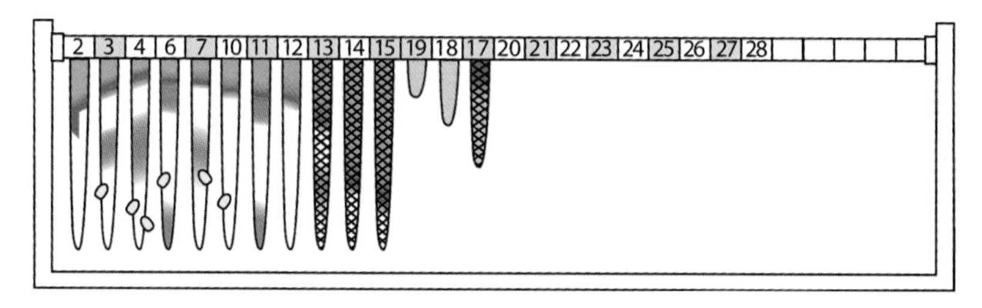

FIGURE 4-12. This illustration shows a hive that is getting ready to swarm. It is full of queen cells that are close to ripening. The queen, along with five combs of capped brood and honey (the minimum size for a divide), is moved to a new box. The combs being moved are checked thoroughly in order to remove any active swarm cells. The old hive is left with multiple swarm cells so that it can raise a new queen and is spaced slightly to make room for new growth. The new divide will have lost its field force, so its growth will remain slow until the capped brood hatches and the young bees graduate into foragers.

- Uncapped Brood
- Ripe Honey
- Unripe Honey
- Crystallized Honey
- Pollen
- Empty Comb
- Capped Brood
- Old Black Comb
- Drone Comb

Removing Old Comb

The next level of management involves preventing disease. There are times when disease is unavoidable and a genetic weakness in the bees simply is being expressed. However, there are management practices that can help to keep the hive clean and free of mites, wax moths, and larval diseases. The most fundamental of these is to cull old black combs.

When larvae are raised in a beeswax cell, a sheer, less-than-paper-thin cocoon is left behind. After hatching, the rough edges are chewed away, the shell of the old cocoon is varnished with bits of propolis and beeswax, and a new egg is laid inside. Once the egg hatches, another larva spins its cocoon and leaves it behind when it is finished pupating, and this process is continued again and again. As the cocoons accumulate inside the original beeswax cell, the thickness of the wall gradually increases and the diameter of the cell gradually decreases. Trapped beneath each cocoon layer is a bit of larval defecation that is sealed inside the cell. Although honeybees try to keep themselves clean, they have bits of oil and propolis on their feet and leave a travel stain as they walk over the comb. Eventually, the combs begin to turn from a creamy white to black. The layers of cocoons become an area in which bacteria, fungus, and wax moth larvae can proliferate. These combs eventually become a liability to the colony.[1]

Dr. Elbert Jaycox was a bee research scientist at the University of Illinois who researched the effects of old black combs in beehives. He found that the old combs darkened the honey and wax produced in them and increased the incidence of brood diseases.[2] He recommended that beekeepers cull broodcombs every three to five years, or whenever they could no longer see light shining through them when holding them up to the sun. Culling old comb is routine maintenance for us, and we do so by rotating the dark comb to the back of the hive, where it can be filled with honey and harvested away. Occasionally, throughout the season, as fresh comb is being built, it is necessary to shift one or two fresh white combs to the front of the hive, slowly shifting all of the older combs toward the back.

Honeycomb is a tremendous resource for the hive, so we don't discard empty comb unless it is showing signs of disease already. The bees must

FIGURE 4-13. A pile of old black comb ready to be melted down.

consume 9 to 11 pounds (4–5 kg) of honey in order to excrete 1 pound (0.45 kg) of wax, so comb shouldn't be discarded lightly. In the springtime, when there is an abundance of empty comb, we spend time evaluating the combs and rearranging them for the summer season. Rather than discarding the older combs, we allow them to serve one last purpose for the hive by becoming harvestable honeycomb. If this sort of maintenance is done each season, there never will be a buildup of dark comb in the hive.

Beyond honey harvesting and disease prevention, there are many other ways of managing the hive; swarm suppression, queen-raising, and winter hive maintenance are just some of them. Ideally, a beekeeper forms a healthy symbiotic relationship with the hive over time, such that both the hive and the beekeeper benefit from the relationship. A positive codependency provides the beekeeper not only with honey but also with a wonderfully therapeutic hobby. Those of us who spend a lot of time inside beehives can testify to the amazingly peaceful and radiant energy that emanates from a well-maintained hive. It is truly a blessing to become a welcome participant in the microcosmic world that honeybees inhabit.

Killing Bees

Part of our ethical commitment as beekeepers is to avoid killing bees when we work our hives. It is quite possible to move through a hive, harvesting honey or doing routine maintenance, without killing bees. The beekeeper must remain calm and patient while using techniques that prevent smashing bees or disrupting the hive. Of course, there are times when bees will be killed, but we do our best to prevent this. Here are some simple techniques to help avoid upsetting a hive or killing bees:

- When using smoke, make sure the smoke is cool. If flames are spouting from your smoker when it is puffed, it is not ready and needs more fuel placed on top of the burning material.
- Don't use too much smoke. A few generous puffs at the entrance (not *in* the entrance) and a couple of puffs at the back when the top bars are lifted is plenty. Only if the bees are noticeably cranky should more smoke be applied. Smoke should never be puffed directly at bees at short range. They don't have eyelids, and this will confuse and upset them, causing an alarm to be set off within the hive. If the right amount of smoke has been used, the bees will busy themselves eating and largely will ignore the beekeeper.
- Move slowly and deliberately through the hive. Sudden movements over or within the hive will set off an alarm. Bees move at such a high frequency that they cannot perceive slow motion. They notice changes in light, however, so whenever possible work through a hive when the sun is high in the sky, and take care not to cast shadows over the hive.
- Become comfortable with using soft weeds to brush bees from unwanted locations. Hive tools or other synthetic tools are not very helpful for this purpose. A bunch of soft flowers, grass, or tree leaves work very well for this purpose and should be wielded gently and

often. Brush with small circular strokes, lifting the bees up and off the comb. The bees should be brushed into the open hive rather than onto the ground, where they can find their way onto feet and up pants legs.
- When putting top bars back together, don't crush the bees between the top bars. Just gently bump the bees out of the way. Bees are continuously bumping into one another within the hive, and some gentle nudges usually will suffice and encourage them to move. Lifting the bar slightly and then bringing it gently down over the bees that are peeping upward can help with this. Lightly smoking the gap between bars also can help.
- Leave the hive safely secured when the work is done. Bees prefer their hive to be airtight and will go to great lengths to fill holes with propolis. If the hive is left unsecured each time it is worked, the bees will have extra work to do to restore the airflow and circulation that is so important to the stasis of the hive. We keep many thicknesses of wood strips (left over from cutting top bars) to put at the back of the hive to secure the top bars tightly against one another. As the hive expands and contracts, this last strip of wood may need to be changed out time and again for another of a different thickness. Sometimes during an extreme heat wave, when the bees begin bearding heavily on the outside of the hive to stay cool, we will leave this last strip of wood removed, so that the bees can circulate fresh air into the back of the hive.
- Carry respect, reverence, and love into all interactions within the hive. Bees can be profound mirrors for the deepest parts of ourselves. Coming to a hive in a bad or irritable mood can prove to be painful.
- Overcome the fear of being stung. Fear is an impetus for irrational and impulsive behavior. A good beekeeper must be willing to take the pain that the bees will sometimes offer and learn from it, so that stings happen less and less often.

We hear from many new beekeepers that they do not want to disturb their bees, so they rarely look inside their hives. We encourage them to spend more time with their bees, forming an alliance that is comfortable and effortless. Although it is best to avoid killing bees, a fear of harming the bees should not deter a beekeeper from interacting with the hive. The long-term goal of any good top-bar beekeeper should be to feel welcome within the hive and to feel an exchange of energy that is positive and sustaining all around.

To Feed or Not to Feed

In a natural beekeeping system, honeybees eat what nature provides them: honey and pollen from flowers. The hive is managed to support the natural fluctuations in nectar and pollen flows, and unless the beekeepers interfere by taking too much honey for themselves, a colony usually can support itself in most climatic conditions. There can be times when emergencies occur and a little sugar syrup or soy-based pollen substitute can keep them alive, but in an organic system it is considered an undesirable last resort rather than a regular replacement. We do not feed our bees unless they are starving, at which point we feed them our own crystallized honey, and, when we no longer have honey, we purchase honey from a beekeeper who we know produces honey without the use of antibiotics or other chemical inputs.

It is very rare that an established hive needs feeding, as long as the beekeeper is conscientious about leaving the hive with enough of its own resources to carry it through winter and periods of nectar dearth. In the last fifteen years, we have fed our bees only a few times, most often in late spring when the resources left over from winter have been exhausted and the spring bloom has frozen due to a late frost. This is a difficult period of time for beehives because they are focused on expanding due to the warm, lengthening days, and require good reserves of nectar and pollen to do so. If the initial nectar flow abruptly ceases and the hive has used up its winter reserves already, it suddenly has too many mouths to feed, causing a starvation crisis.

Standard recommendations within the beekeeping industry are never to feed bees honey because it can carry disease. Instead, bees are fed sugar syrup or high-fructose corn syrup solutions. Because these products are pasteurized, highly refined, and do not carry any potential disease-causing fungi or bacteria, they are considered safer for the bees. This recommendation originated with the fear of spreading foulbrood spores in the honey. Although it is meant to protect honeybee health, feeding bees sugar syrup also has proved to be economically smart because it costs less than honey, can artificially prop up hives that would otherwise be lost, and boosts the size of hives during nectar dearths.

Honey is, without question, the best food for bees. Honeybees gather and distill it for their own consumption, and it is a much more complex and complete carbohydrate than sugar syrup. If the hive is in a healthy condition and has disease-resistant bees, then any potential disease-causing spores or bacteria should not be a problem. Similarly, human beings continuously are exposed to all kinds of potentially harmful bacteria and fungi, yet our immune systems are designed to resist these disease-causing agents, as long as we are in good health. Diet is often the basis for good health, not only for us, but for honeybees as well. Unselfishly allowing bees to eat the honey and the pollen that they have gathered for themselves is a beekeeper's best defense against hive disease.

Crystallized raw honey is easy to feed because it can simply be scooped onto the back floor of the hive. In the absence of crystallized honey, liquid honey or sugar syrup can be put into a vacuum feeder and placed inside the hive. To make a vacuum feeder, take a jar that is small enough to fit under the bars of the hive, punch ten to fifteen 1⁄32- to 1⁄8-inch (1–3 mm) holes into the lid, and turn the jar upside-down over two sticks that are ½ inch (1.25 cm) in diameter. The bees can then crawl between the sticks and drink the liquid until it is gone.

Large commercial operations have been feeding high-fructose corn syrup to bees because it is widely available and inexpensive. Recent research, however, has shown that a compound in high-fructose corn syrup can actually kill bees.[3] Dr. Diana Sammataro at the Carl Hayden Bee Research Facility placed some beehives in a large greenhouse where the bees had no natural flower-based food. The only food available to them was high-fructose corn syrup and pollen substitute. The hives died of starvation even though their combs were full of the syrup. Her hypothesis was that these substitutes were indigestible to the bees.[4] If beekeepers are in need of a quick fix to feed their bees, white sugar syrup may be used as a substitute for honey, but high-fructose corn syrup should be avoided.

It is important not to feed honeybees heated honey (such as the honey left over from processing wax), brown sugar, or molasses. These products have been shown to contain impurities that cause a sort of dysentery in the bees, creating symptoms similar to nosema infection. We allow our bees to clean the honey from our wax before it is melted, by placing it outdoors in open containers where the bees will sort through it, licking the remains of honey off of the wax surfaces. Once it has been cleaned by the bees, we then melt the wax.

Honeybees gather and store pollen in an amazingly complex and careful fashion. They gather this protein-rich food by rubbing their thorax fur on the anthers of flowers, then comb the pollen off of their body with special pollen combs on their middle and front legs. They then pack it into pellets around bristles on their back legs and carry it back to the hive. In the hive, pollen-carrying bees look for cells near the brood nest in which to store the pollen. Then they lower their back legs into these cells and knock off the pollen pellets, sometimes layering many types and colors of pollen into one cell.

Honeybees thrive on the variety of natural pollens stored in the combs. Just as they turn nectar into honey by mixing it with enzymes and beneficial bacteria, they also inoculate the pollen with special fungi that break down the pollen coating and make the protein and fatty acids more available for digestion.[5] This fungal inoculation also prevents the lipids from going rancid and preserves the pollen for future use. This process is similar to the one humans use in making cheese.

Pollen that honeybees have stored is nutritious and easily digested, and has been adequately preserved. Neither honeybees nor humans can digest the uncured proteins in raw pollen nearly as well as they do this carefully prepared product.

Bees can extract some nutrition from protein sources other than pollen. They sometimes gather grain dust as a substitute for pollen when there are no flowers blooming, presenting a curious presence at chicken feeders and compost piles.

There are many protein-rich, soy- and grain-flour-based formulations available on the beekeeping market that try to mimic the nutritional quality of the pollen from flowers, but none of them is a true substitute for the real thing. If bees have no pollen available to them, these substitutes may save a hive from starvation, but their regular use is not recommended.

In my more than thirty years of keeping bees, and in my many visits with beekeepers around the world, I have never seen a hive starve to death from a pollen shortage. These supplements are used mainly by commercial beekeepers to create artificial growth during times when natural conditions

FIGURE 4-14. Dandelion pollen is reputed to have a very nutritious and healing quality for honeybee hives coming out of winter.

would not support growth, or in environments in which monoculture predominates and there isn't enough natural plant life to support honeybee populations. Feeding bees sugar syrup and pollen substitutes in environments in which they otherwise could not survive is simply unsustainable and has questionable impacts on the long-term health of honeybee populations.

Although we strongly advocate against using food supplements with honeybees, we try not to become dogmatic about our beliefs. When it comes to a situation in which a hive faces starvation, especially a newly hived package or nucleus hive, it is difficult to make an argument for this kind of rigidity. Bees love sweet things and will appreciate the kind help provided by the beekeeper in times of dearth. We are careful to encourage young and new beekeepers to follow organic practices without making their idealism a stumbling block.

As an alternative to providing these kinds of artificial supplements, we actively encourage beekeepers and landowners everywhere to reinvigorate their natural surroundings by planting a diversity of pollinator forage species. There's a large selection of annual and perennial cover crops as well as ornamental plants, trees, and shrubs that can provide bees and other beneficial insects with an almost continuous supply of nectar and pollen in any given location.[6] There are even plants that provide nectar and pollen in winter conditions, and plants that bloom during droughts or in desertlike environments. Although it is not a quick fix, planting beeyard sites will provide the honeybees with much better nutrition in the long run, and it's an excellent step toward creating a more sustainable world for honeybees to live in.

Many of the plant species that honeybees adore are weedy species like dandelions, which don't require a lot of effort or expense. The last chapter of this book provides a list of plants that are good food for bees and that also can create a rich and beautiful environment for the beekeeper.

Despite all the effort we put into propping up our industrial systems with products such as sugar syrups and pollen substitutes, we cannot possibly compete with the depth and complexity of nature's innate wealth and wisdom. Rather than fighting the natural world, we need to become willing participants in its revival. It is one very active and positive step we can take toward saving the honeybee.

FIGURE 4-15. On our organic farm, we plant blooming borders for all of the pollinators. These calendula flowers are particularly well loved by native bees and syrphid flies. They also can be harvested and dried for use in a healing beeswax salve.

CHAPTER 5

The Seasons

Although we've worked with honeybees as far away as Azerbaijan, Nicaragua, Ecuador, and Mexico, the vast bulk of our beekeeping experience is in New Mexico. We occasionally mention times of the year when bees are swarming or getting ready for winter. Here in New Mexico, the flowers begin to bloom in March and April. Honey gets stored from May through September, and October is our fall, when bees are getting ready for the winter.

In other places in the world, bees make honey in the cool of winter because the intense heat of summer dries up the floral resources, forcing the bees into a heat dormancy when they rarely leave the hive except to gather water. In tropical areas the bees might make honey in the wet season, as in much of Africa, or in the dry season, as in tropical Asia and South America.

Established beekeepers are the best resource for finding out what the bloom patterns are like in any given area, and beekeeping associations are a great way to trade stories and find mentorships. Many beekeeping groups have publications that list the native plants that provide nectar for bees, and offer other valuable resources for beekeepers who are just getting started.

Spring

When a hive comes out of winter, it is a tight cluster of bees around a queen who is laying very few eggs. As the days get longer and warmer, the queen is spurred to build the hive's population by increasing the number of eggs that she lays. When the hive population expands, there are more workers to gather nectar and pollen as the plants begin to bloom.

Choosing the race of bee that is best in a given area can be an important choice when it comes to overwintering and spring growth within a hive. There are some races of bees that will build quickly in the springtime, only to starve to death in a sudden cold snap when, contrary to their expectations, the nectar dries up. Other races of bees build more slowly and may make less honey, but will be hardier during cold weather. Again, established beekeepers are a valuable resource in discovering which races of bees are best adapted to any given area.

Early spring is the time when a hive is most likely to starve and the beekeeper needs to be most diligent in checking the honey stores left over from winter. If there are a number of empty combs between the brood nest and the remaining honey, it is best to move the empty combs toward the back of the hive and bring the honey closer to the brood. If a hive is completely out of honey and there is no nectar flow going on, then the hive must be fed. Good planning going into winter usually avoids this problem, but there is always the chance that spring will arrive later than usual or that drought will occur and the nectar just doesn't flow.

Because most of the combs have been emptied of honey and pollen during the winter, spring is an excellent time to clean and rearrange the brood nest. Our spring cleaning consists of evaluating each comb to see where best to place it in the hive for the new season. If a comb is old and dark-colored but still usable, we move it to the very back of the nest, where it can later be filled with honey and culled. If a comb is made mostly of drone-sized cells, we do the same. Drone comb is not useful during the early expansion phase of the colony, as the focus will be on expanding the worker force and eggs will not be laid in drone-sized cells. Drone comb is best moved to the back of the hive, where it can be filled with honey or with drones once the hive has adequately expanded and has begun to prepare for swarm season. Any combs still filled with honey from the season before are moved to the entrance, where the bees will continue to feed on them as they expand the nest.

If the spring is a good one and the fruit bloom comes on strong, the hive will begin to expand rapidly and will need to be checked often because the influx of nectar and pollen may spur the hive to propagate itself, causing a swarm.

Swarming

Swarming is the reproduction of honeybees on the colony level. It is similar to cell division because one colony becomes two. The colony begins to raise a new queen and sends off its old queen in a swarm with 3,000 to 5,000 young bees filled with honey and pollen. The swarm will gather somewhere, often on a tree branch, and send out bees to scout for a new space in which to live. Once they find a new home, they rapidly begin to use their stored honey and pollen resources to create wax combs in which to live and breed.

Beekeepers have long noticed that careful manipulations of the hive can either discourage or encourage swarming. There are times of the year when bees have a particular desire to swarm, especially in the midst of a strong nectar flow. Although springtime is the characteristic time for swarming, it can happen even in the fall. When the bees are limited to a small space, such that nectar builds up in the brood nest, the hive will be inspired to swarm, regardless of the time of year. Forcing this condition to occur in a cell-building hive is how queen breeders initiate the queen-rearing process.

If a hive casts a swarm, there is a possibility that it will end up being a nuisance to neighbors and that the beekeeper will lose the bees. If, however, a beekeeper rearranges combs inside the hive in the springtime to open up the brood nest, thereby providing space for the bees to build new comb, the swarming urge will be diminished. Thus, swarm suppression, or stretching the hive, becomes one of the main focuses of spring and summer hive management.

At the end of winter, there are a number of empty combs near the entrance to the hive box. The bees have consumed all of the honey and pollen in them and have moved their cluster toward the center of the hive. They will have a small brood nest there and will be eating the crystallized honey at the back of the nest. As the nectar flow resumes in the spring, the hive will want to increase its population as quickly as possible to take advantage of all of the resources that summer has to offer. More combs will need to be built to house the brood expansion, but the crystallized honey (if there is any left after winter) makes a barrier of cold, full comb that cannot accommodate

FIGURE 5-1. Swarm cells often are located on the edges of the combs. This picture shows one fully capped queen cell and one in progress.

brood. If left trapped in a space that is too small for their needs, the hive will begin to raise queens in order to swarm.

By simply moving the honey-filled combs to the doorway and the empty combs to the back of the hive, the beekeeper can suppress the swarming urge at the onset of spring. The brood nest is left undisturbed and the bees can now begin to expand toward the back of the hive. The beekeeper has opened up the back of the hive effectively and will continue to stretch the nest by using the spacing techniques described in the Hive Management section of the text (see Chapter 4). The expansion of the nest serves to redirect the energy of the hive into comb building, brood development, and nectar gathering, rather than swarming. Of course, there is only so much space that can be made, and the swarm urge may reappear readily once the hive is full. At that time, making divides is the best option.

The vigor of a queen greatly influences the swarming urge in beehives. Hives with old or poor queens tend to be more interested in sending off the old queen in a swarm and raising themselves a strong new queen. Hives with a young vigorous queen that is exuding ample queen pheromones are more likely to put off swarming.

Some beekeepers clip the wings of queens as a method of preventing swarming. The idea behind this method is that the queen's inability to fly will keep a swarm from happening. If the hive is congested, the bees nonetheless will raise a new queen and try to send off the old queen, whether she can fly or not. The swarm will then end up queenless, and the old queen will be kicked out of the hive or killed. We do not recommend this method of swarm prevention and don't clip wings for any reason. If a hive has made up its mind to swarm and has laid eggs into queen cups, we follow its lead and make a divide.

Making Divides

Dividing hives is the best way to propagate bees and increase the number of hives in a yard. It usually is done in the springtime, when the natural instinct to swarm is peaking. In years when there is a strong nectar flow throughout the summer, the swarming tendency can be found as long as the nectar flow persists, and successful divides can be made late into the season. However, as the season progresses, a greater number of combs must be provided to new divides in order to bolster their ability to make it through winter.

Although it is possible to make divides when the swarming tendency is not present in a hive, studies have shown that a hive will raise stronger queens when the swarming impulse is active. Forcing a hive to raise queens when nectar and pollen are scarce has a lower chance of success and generally will yield weaker, smaller queens.

In the spring, as the first nectar and pollen flows of the season start coming into the hive, the queen will be stimulated by the workers to increase her egg laying and the brood nest will begin to expand. In the first stage of expansion, the hive will raise lots of workers and then, as the workforce reaches a critical mass, they will begin to build drone cells and direct the queen to lay unfertilized drone eggs. This is a sign that the hive is getting ready to propagate itself.

Some hives expand more quickly than others, and usually the queen's laying pattern is stronger and more uniform in these hives. If they are also docile bees and produced a good amount of honey in the previous season, they are the best choice for making divisions. Choosing these strong and gentle hives to propagate from is a form of genetic selection. Over time, the strength of the genetic pool will grow as a result of selecting for queen vigor and other important genetic traits. Keeping good notes on each hive can help to facilitate this process.

Before making a divide, check for the following conditions to see if it is the best time to do so.

Is there a strong nectar and pollen flow?

In springtime, there are often good flows during fruit-tree bloom, but it is possible for the nectar flow to dry up after fruit bloom. Look around carefully to see if there is a good amount of flowers in the area for the bees to forage from. There should be ample wet nectar and pollen in the combs to support the kind of effort that the hive will have to make in order to raise new queens.

Are there swarm cells being built and frosted at the edges of the combs?

When the hive begins to feel that it is time to swarm, it will begin to build queen cells in preparation for raising new queens. Queen *cups* usually are

FIGURE 5-2. This photo shows an advanced queen cell with a large larva curled inside.

found at the edges of the combs. When the bees begin to polish the interior of these cups and *frost* the outside with fresh wax, their attention is starting to shift toward swarming. The hive should be checked at least once every two weeks during swarm season so that the bees can be divided as they begin this process. As soon as the bees begin activating queen cells by frosting them and directing the queen to lay eggs in them, the hive should be divided to take advantage of this momentum.

Are there eggs and royal jelly in any of the swarm cells?

The ideal time to make a divide is when there are eggs in the swarm cells but the larvae are not yet very developed. At this point in the swarming process, the bees have little invested in the change they are about to create and will accept the rearrangement of the hive more amicably than when the queens are more fully developed. We've certainly made many divides when the queens were almost about to hatch, but there is a greater possibility of something going wrong, so it's best to catch it early.

Is the queen strong?

When a divide is made, the queen is taken away from an abundance of workers and put into a situation in which she has lost her field force and has mostly nurse bees to support her. If she is a strong queen, this will not be a problem, and she soon will have laid a whole new array of workers to join her. If she is not a strong queen, the divide may never get off the ground.

If the hive has a weak queen, there may actually be *supersedure cells* (these are found in the body of the comb, whereas swarm cells are usually on the edges of the comb) rather than swarm cells present. In this case, the best strategy is to kill the weak queen and allow the bees to raise a new queen instead of making a divide. If a daughter of the weak queen is undesirable because the genetics of the hive aren't strong, a swarm cell from a genetically desirable hive also can be inserted to replace the supercedure cells. Although it may seem heartless to kill the old queen, it can be beneficial to do so because the bees that might have followed her out in a swarm when the new queen was ready to hatch will be retained. Thus, the hive will be stronger when the new queen arrives.

Is the hive large enough to divide?

There need to be at least eight combs of bees in a hive before it can be divided. This doesn't include empty or unoccupied combs. The smallest nucleus hive we can recommend is four combs and, more often than not, we try to make six- or eight-comb divides. As the season progresses, we make larger and larger divides to ensure that both the old and the new hives have enough resources for winter. A four-comb divide is recommended only in early spring when there is still ample time for the bees to build the twelve or more combs they will need to get them through winter.

If the time is ripe for making divides, these steps can be followed to make a successful divide.

Find the queen and move her to a new hive.

Ideally, the queen would be moved to a new yard altogether, but this is not absolutely essential. If she is being kept in the same yard, though, it's best to move her as far away from her original location as possible.

Check the comb with the queen carefully to make sure there are not any swarm cells with eggs in them. If there are, they may be broken down or the queen may be moved gently to a new comb that is free of them. We don't break down any queen cells with eggs if we can avoid it because they are a valuable resource. We distribute them among hives that are queenless or have a failing queen, and we make sure to leave at least one behind for the hive that is being divided.

When moving the queen from her present hive into a new box, it is best to lay the new box close by. The short distance helps to ensure that she doesn't fall off the comb in transit.

So how do we find one special bee among tens of thousands? Practice and patience help. In the spring, when there are fewer bees, it is easier to get the hang of it. We look over the comb for anomalies and try to avoid getting distracted by individual bees. The queen is bigger and walks with a swagger, like a pregnant woman, often leaving a wake of empty space behind her. If a minimum of smoke is used, the queen often will be in the brood nest, sometimes even still laying eggs in the cells as we search. (It can be extra-hard to spot her if her abdomen is buried in a cell laying an egg.) Sometimes the queen will try to hide from disturbance and will be running around on the bottom or in the back of the hive, but usually she can be found walking around on the combs.

Give the new hive combs that are primarily capped brood.

Once the queen is safely in a new box, look through the remaining combs in the hive and choose combs that are free of swarm cells and have mostly capped brood to transfer to the new hive. This queen is losing her field force, so she will benefit most from broodcombs that are on the verge of hatching and no longer need to be fed. If she is given an abundance of eggs, some of this brood may perish due to a lack of bees to care for it. If she is given bees on the verge of hatching, the hive will fill more quickly with workers who can forage.

FIGURE 5-3. Dark-colored queens are often harder to spot, but this beautiful queen shows up because of the wake behind her and the light color of the wax. Her laying pattern is nice and tight – a comforting vision of hive health.

The original hive will be queenless and will not have to care for any new eggs, so they can bring the eggs and young larvae more easily to maturity. If there is a strong nectar flow, the hive also may fill quickly with honeycomb because the workers won't have any fresh brood to care for until the new queen is laying and can dedicate their time to foraging.

Give the queen-right divide some combs of honey and pollen.

Transferring honeycomb to the new divide will provide them with adequate resources while they get off the ground. There will be little foraging force in the first few days of the divide because most of the older mature bees will return to their original hive. The nurse bees will need honey inside the new hive to take care of the queen and the brood until they have matured and are ready to start foraging. Put the honeycombs in front of and behind the broodcombs and keep the brood together so it can be kept warm.

Brush some additional bees into the queen-right hive.

Once enough combs of brood and honey have been moved into this new divide, gently brush two more combs of bees into the new hive and then take the unoccupied combs back to the old hive. This is done in order to get more baby bees into the new hive. Although four or five combs were moved into the new hive, the older mature bees will end up back in their original hive once they go outside to forage. Brushing in extra bees ensures that there will be enough nurse bees to support the brood that was moved into the new divide.

Arrange the original hive for queen-raising.

Check the original hive to ensure that the combs have a swarm cell with an egg in it. Ideally, the divide was made because the hive was already on the verge of swarming and had a queen cell in process. If they didn't have this kind of resource and the divide is being made out of season, then lifting a young larva from another cell and putting it into a queen cup helps a great deal to start the process. This will provide the hive with a better chance of raising a new queen and will help to avoid the possibility of laying workers. Chapter 9 on Raising Queens will help to explain how to do this.

The more larvae that are transferred into swarm cells, the greater the chance that the hive will be successful in raising a healthy new queen. If young larvae aren't transferred, the bees most likely will go ahead and raise a queen from a worker cell with an egg, but the quality of the queen may be somewhat compromised by the smaller cell size, and the hive may even end up with an intercaste queen. Intercaste queens are born of female eggs that were being raised as workers and then were modified to become queens too late in the process, so that they will never become viable queens. (See the Intercaste Queens section on page 113.) This can happen if the bees don't start raising the queen soon enough after the divide. With a little bit of practice, lifting young larvae (less than two days old) into swarm cells can be an easy and quick way of ensuring the success of a divide.

Once the divide is complete, it can be moved to its new location in the yard and monitored regularly to see that the queen has adjusted to her new arrangement and that the hive is growing at a steady rate.

It is best to leave the original hive alone until its queen-raising process is complete. Marking the calendar with the date of the divide helps you to know when the hive needs to be checked again. Depending on whether there was an egg in a queen cup at the time of the divide or whether there were capped queen cells already, the wait may be as little as two weeks or as long as a month before there is a new queen who is mated and laying.

After the divide is complete and a significant amount of time has passed, it is possible to go into the hive that was raising a queen and still not see a mated queen. In this case, we look for signs of a queen's presence. There should be lots of cleared space in the combs, with several contiguous combs in the busiest part of the hive that have a circular or oval area that is empty, clean, and ready for a queen to lay eggs in. If there are combs full of honey or nectar and pollen, with no cleared space at all, the queen-raising was not successful and the hive is queenless. If there is lots of cleared space, it may be that she is not yet laying or is out on a mating flight.

If we check a hive that was raising a queen and don't see signs of a laying queen yet, we bring in a comb of eggs from another hive, just in case they need another chance. It could be that there is a queen present, unmated or not yet laying, but this comb of eggs is a deterrent against the possibility of laying workers and can help the hive to feel settled until their queen is in action. The brood exudes hormones that suppress worker-bee ovaries from developing into laying workers. It helps to note which bar the donated eggs were on and to check again within a week to see if the status of the hive has changed at all. If they weren't successful with their original queen-raising, hopefully they will have started to raise a queen from the comb of eggs that was recently donated. Sometimes hives are simply unsuccessful in raising a queen, and then they become candidates for combining with other hives.

Often in the springtime, when a hive is feeling strong and there is a good nectar flow, there will be multiple queen cells in a single hive as they prepare to swarm. This is an opportunity to make multiple divides from one hive if you have enough bees and combs to support the divisions.

We also routinely donate extra queen cells to hives that are weak, after first killing their old queen. It is best to use a queen cell that is not yet capped for this purpose, as the hive may be less likely to accept a more mature queen cell so soon after their queen's death. With that said, we've certainly put mature queen cells into weak hives after killing their queen, and have had great success requeening the hive. If a mature queen cell is the only resource available, it is probably worth a try.

Swarm season is a tremendous opportunity to strengthen and multiply hives, and over time we've come to see that a swarm lost is an opportunity lost. We make every effort to work our bees intensively during this time so that we don't lose any swarms and can make selections that will strengthen our genetic pool. It is the most exciting and active time of the season, as the potential energy of new queens is harnessed and nurtured into fruition. Even the thrill of a hive full of honey doesn't match the sight of a large and beautiful new queen taking her place in nature.

The Two-Queen System

During the winter the bees occupy only twelve to fourteen combs. We often will overwinter our hives in double boxes, which are made from a top-bar hive with a partition down the middle and an entrance at either end. The two-queen system is a way to take advantage of this arrangement in the springtime in order to make honey and divides simultaneously.

When two hives occupy a partitioned box, they run out of room quickly in the spring because there is little room for expansion. Once the nectar flow begins in earnest, the limited space induces one or both of the hives to begin raising queens in preparation for swarming. Once a new queen is on the way, a six-comb divide can be made from each hive. This removes the queens from both sides of the box, along with their attendant bees.

Once the queen-right divides have been removed from the partitioned box, the partition can be removed and the two clusters of remaining combs can be combined. Very little fighting will take place because the queens have been removed and a new queen has yet to spread her pheromone throughout the hive. Soon they will have a new queen, but there will be a period of lag time during which she will have to hatch, mate, and then begin laying. This means that the field bees that are left in this hive won't have any nursing to do for almost a month and can dedicate their time exclusively to gathering nectar. This newly combined hive will have approximately sixteen combs and two foraging forces of workers who are already familiar with the flowers that are currently blooming.

If this procedure is done during a good nectar flow, the hive will fill with harvestable honey, which can be removed just as the queen begins to develop her brood nest. This is an excellent way to obtain spring honey.

Summer

Summer is the time when everything is happening in the hive. The colony expands exponentially and, if the weather is good, it fills with brood and honey. Life is good for the bees and the beekeeper. There is also a lot of work to do to keep up with the pace of the hive: spacing, harvesting, evaluating queens, making divides, monitoring for disease, refining wax, and bottling honey. It can be a fast and furious blur of activity if one is making a career of beekeeping. Even if beekeeping is just a hobby, the tasks at hand still can get away from you unless some diligence is maintained. Often beekeepers will wait until the end of the season to harvest honey, not realizing that the hive may fill up and swarm late in the season, potentially leaving the hive weak or queenless in the fall. Honey harvesting sometimes can be like deadheading blossoms: the more you harvest, the more the bees will make to harvest. When the nectar is flowing, the beehives should be checked at least once every two weeks to keep them in good shape.

FIGURE 5-4. A freshly capped honeycomb against a background of summer's signature sunflowers.

Bearded Hives

When it is warm and there are flowers blooming, worker bees begin hanging out in masses around entrances or holes in the hive, especially in the evenings. Often this phenomenon makes the beekeeper worry that the colony has grown too big for the hive and that the bees are about to swarm. However, if there is still space in the hive to build new combs and no sign of queen cells, there is nothing to worry about. The bees are hanging outside the hive so that their body heat and humidity will not add to the heat and humidity in the interior of the hive. It is more difficult for the colony to ripen honey when heat and humidity are too high. Bearded hives are generally happy hives, and bearding is not necessarily a sign of imminent swarming.

Mixed-Origin Divides

In the middle of the honey-making season, after the swarming urge generally has diminished, it is possible to make a few new hives in another fashion. Rather than pulling a whole divide from one hive, we instead take two or three combs from four or more hives. This doesn't weaken the donor hives in any appreciable way, and it ensures that the new hive will be large enough to make it through winter. Ordinarily, bees from different hives would fight with one another. In this case, there are bees from so many different hives that they lose their orientation altogether, and very little fighting takes place. It is very important *not* to put a queen into a mixed-origin divide. We want this new hive to be totally free of any allegiances so that the bees will interact peacefully with one another.

This is a good way to use an extra queen cell or an unaffiliated caged queen during the middle of the season. However, a mixed-origin divide should not be considered a good candidate for raising its own queen from eggs. It will lose any field bees to the donor colonies, so it should be given mostly capped brood and three or four combs already filled with honey and pollen. The mixture of bees will raise the queen cell to fruition or release their new queen from her cage and form their own hybrid hive.

Fall and Winter

Once summertime is over and the nights begin to cool, the colony will begin to contract its brood nest. In the fall, the queen may cease laying eggs completely for a short period of time to allow this diminishment to occur. When she begins laying again, it will be just enough eggs to keep the hive population constant during the deep of winter. The relatively small cluster of bees will migrate slowly through the honey and pollen that were stored during the summer, keeping themselves warm through the vibration of their wings.

In preparation for winter, the colony will start to propolize the hive entrance and any other cracks or crevices in the hive body in order to control the air circulation and temperature in the hive throughout winter. They will reduce their hive entrance to a small fraction of its summer size, not only to control the air quality within the hive but also to fend off predators such as mice more easily, predators who are interested in eating their honey and pollen reserves.

Throughout most of the southwestern United States, we have found that twelve combs support a colony over the winter with plenty of food for survival until spring. In warm climates, where a period of dormancy is due to excessive heat, bees consume a lot of honey and pollen because they stay active and gather water to cool the hive. However, in a climate with cold winters, bees become much less active, expending as little energy as possible and seldom leaving their tight cluster. When experimenting to find out what works best in a particular climate, leaving more reserves than necessary is better than taking too much and finding that the hive starves before the nectar begins to flow again.

The end of the honey flow is not the time to rearrange combs or take pollen combs because the bees have arranged the honey and pollen carefully so that the winter cluster will be able to eat a winding path through the combs slowly all winter long. If it is really cold and the cluster comes to a comb with no pollen, they are unable to move to another comb to retrieve pollen because any bee that leaves the cluster will freeze. They could be inches away from food and yet still starve. All the preparation for winter is done while the hive is still being harvested. Drone combs and older darkened combs are moved to the back and culled before the end of the season, leaving the first twelve combs in the hive relatively undisturbed.

Some hives may have built only six or seven combs throughout the season and will not have sufficient reserves to survive the long period of dormancy. Beekeepers have a couple of options in trying to save a hive like this. Either the hive can be given combs of honey and pollen to boost its reserves or it can be combined with another hive that is similarly too small to endure winter.

Combining Hives

Combining hives can become a routine part of fall hive maintenance. When combining two honeybee colonies, it is best to choose the stronger of the two as the receiving hive. The receiving hive has an advantage because it retains its queen and its field force, while the weaker hive will have to move to a new location and submit to the will of the bees there. The queen in the weak hive must be removed or the two queens will fight. If she cannot be found, the hives may choose one or the other of the queens successfully, but they also may lose both in the process, so it is best to find her and remove her before moving all of the combs.

There are a couple of ways to combine hives. One is to drape a sheet of moist newspaper over an empty top bar at the back of the receiving hive's combs. The edges of the newspaper are tucked against the walls of the hive, forming a barrier from one hive to the other. This prevents the hives from engaging in an immediate war after they've been combined. A tiny tear (½ inch, or 1.25 cm) is made in the newspaper where it contacts the floor of the hive. Then the combs of the weaker colony are added right up against the partition, so that the top bar with the newspaper is the only thing separating them. The bees will smell each other through the paper and begin chewing holes to unite their colonies. By the time they mix freely, they will have become accustomed to each other's scent and will form a single hive with sufficient resources to survive the dearth. The newspaper will have been reduced to a pile of fluff that either can be removed by hand or will slowly be removed by the bees.

Another approach to combining hives is to camouflage the scent of the two hives so that they cannot distinguish themselves from one another. To do this, a mixture of sugar-water is made up and scented with some strong essential oil such as mint, lemongrass, or lavender. The combs of both hives each are misted gently and the weaker hive is put directly into the back of the stronger receiving hive. The bees will comingle and begin licking each other. By the time the smell of the essential oil has worn off, the bees will be united as one hive and will accept one another.

Bears

Bears love to eat larvae, honey, and pollen, and they tend to be found in beeyards in the fall, before hibernation. They are also powerful climbers, making them almost impossible to fence out. An electric fence with a 5,000- to 6,000-volt output has been found to be the most effective way to keep bears out of beehives. There are solar energizers with batteries that work well in places where there are no electrical lines.

If bears are in an area where bees are going to be kept, it is important to set up the bear fence first and bring in the bees only after their protection is in place. It is inconsiderate to the bees and the bears to leave bees unprotected in bear territory. It is said that a bear who has tasted a beehive once can smell hives from up to 5 miles (8 km) away. Once a bear gets a taste of brood and honey, it will come back every night until all the hives have been consumed. There are many bears who have become addicted to beehives and wind up getting shot because they have become a nuisance.

The area in which the electric fence is going to be set up should be cleared of vegetation along the perimeter where the fencing will be placed. Fence posts should be no more than 4 yards (3.7 km) apart, and plastic insulators should be put on them to hold the six or seven strands of electric wire or mesh. The electric wire that runs along the bottom of the fence should be no higher than 6 inches (15 cm) above the ground.

Bears tend to approach food with their noses to the ground, and the low wire shocks them right in the face, which is a powerful deterrent and will keep them out despite the delicious smell inside. In dry, sandy soils, a wire mesh should be laid down and stapled to the ground all around the outside perimeter and wired to the ground posts to make sure the bear is well grounded through its feet when it touches the electric wire.

Hive Insulation

In Eastern Europe and northern Asia, many beekeepers use a horizontal hive, often called the Russian long box hive. It houses the brood nest in the center, and frames are added at either end in the honey-making season. To overwinter these hives, beekeepers cover the hive body with a heavy canvas and place an insulating blanket or straw-stuffed pillow over the canvas and then add yet another waterproof layer on top of it all to shed snow. The ends of the hive are stuffed with straw, and the canvas is hung over the enclosures. This keeps out drafts and absorbs some of the humidity made by the warm breath of the cluster. These steps are taken because of the long winters and accompanying extreme temperatures and wind conditions that exist in these climates.

Often beginning beekeepers wonder about whether or not to insulate their hives during wintertime. Although insulation can be helpful in increasing the overall ambient temperature in the hive, it rarely makes a significant difference. Bees have been shown to overwinter in harsh climates without insulation as long as they have sufficient stores of honey and pollen to fuel their heating energy.[1] We do not choose to insulate our hives at all. Although holding in heat, many insulation methods cause a decrease in ventilation, thus increasing the likelihood of condensation's getting trapped in the combs and causing disease. The bees generate a good deal of heat, which rises to the top bars, where it condenses into moisture. If this moisture cannot evaporate upward, it drips back down on the cluster and into the upwardly tilting cells. This can lead to problems with mold, making the hive more susceptible to disease.

Placing hives in front of a south-facing wall is one of the most effective ways of providing some wind protection and extra warmth. Stacking straw bales on the north side of a hive creates the same kind of protection because the bales will absorb heat from the sun during the day and radiate it back out during the night when temperatures plunge. It is important not to let the bales touch the hive body because they may increase the risk of mold or disease.

In severe climates, we overwinter our hives in double boxes. Top-bar hive bodies can be divided with a partition, making space for two hives to be overwintered next to one another. This provides both colonies with the benefit of each other's warmth. An entrance must be located at either end of the hive so that the two colonies have ingress and egress from separate locations, and the partition must be solidly sealed so that the hives do not have access to one another. We have found this to be a very sustainable way of protecting our hives from cold temperatures.

CHAPTER 6

Honey, Beeswax, and Other Products

When the sun is high in the sky and the season is at its peak, harvesting and processing begin. There are many products that come from the hive, from raw honey to beeswax candles, and each of them must be cared for in careful steps to ensure that the final products are wholesome and well made.

Harvesting Honey

We do not harvest any honey until the hive has built at least twelve full combs of brood, pollen, and honey. Anything beyond the twelfth bar is considered extra in our particular system. This is a number that could be variable depending on where the hive is located and how long the winter or period of dearth extends. The goal is to always leave as many resources as the hive needs in order to survive a long dearth of nectar, no matter what time of year it is. Taking honeycombs from this reserve, with the thought that the hive will be able to rebuild it over the rest of the nectar flow, is not recommended. There is never any guarantee that the nectar flow will make up the difference, and by taking from this reserve the beekeeper is gambling with the life of the hive.

When a beehive is expanding at the beginning of a nectar flow, there may be brood on most combs. If there is a honeycomb that has a small patch of capped brood on it, but is otherwise all capped honey, we sometimes will cut out the section with the brood carefully and set it at the back of the hive so that it still can hatch, while harvesting the rest of the comb. This is rare and is mainly considered an option for honeys that will crystallize quickly if they aren't harvested soon after ripening. Generally, honey is harvested when the brood nest begins to contract. If the hive is really full and there seems to be brood on most combs, some space can be made in the hive by making a mixed-origin divide.

Honey is ripe for harvesting when it is either capped with a wax coating or has been evaporated such that when you press the cells with a finger, the honey does not run but forms thick droplets. Sometimes it is necessary to harvest uncapped honey when the nectar flow has stopped and the cells are only half full, yet the honey is nonetheless ripe and will proceed to crystallize if left unharvested. In humid climates, only a very small percentage (5 to 10 percent) of uncapped honey should be harvested, but in arid climates a larger percentage (10 to 25 percent) is generally safe.

There are times when the compulsion to harvest honey is strong, but the honey simply isn't ripe. It takes approximately eight drops of nectar to make one drop of honey. The bees ripen the nectar into honey by fanning their wings and dehydrating the liquid until it becomes thick or ripe. When the honey isn't ripe, it is runny and drips easily out of the comb. This is referred to as wet honey and should not be harvested because its water content is too high and it can ferment in the jar. In tropical climates, where the humidity is extremely high, the bees are unable to ripen the honey, and the nectar ferments in the hive. In arid climates the bees have to expend a lot less time on nectar dehydration due to natural evaporation.

There are certain plants, such as mesquite or phacelia, that yield honey that is very quick to crystallize, and there are also honeys that are slow to crystallize, and even ones that never do, such as avocado or tupelo. Mesquite honey must be harvested immediately after the nectar flow has ended or it will crystallize in the comb and be impossible to extract. Although it is important to wait until honey is ripened before harvesting, it is also important to harvest combs that are ripe, even if they are uncapped, in order to prevent crystallization within the hive.

Honey that is stored in dark-colored wax that has been used previously by the hive for brood, and thus has cocoons in it, is always expressed from the comb by crushing rather than being used for comb honey. It would be

FIGURE 6-1. Cells of honey that are unripe. The honey is still somewhat liquid and slumps in the cell, forming a droplet. The bees will continue to dehydrate this honey, and when it is ready they will cap it with beeswax to preserve it for future consumption.

chewy and potentially contaminated with debris if eaten in the comb state. Beautiful, freshly drawn capped combs, ones that have never had brood in them but were rather built exclusively for honey, always are reserved for comb honey. They are tender and delicious. We carry an empty hive and top bars with us when we harvest so that we can carefully transport these combs fully intact. The hive is lined with a clean towel or a large trash bag to keep the combs clean and to contain drips.

All the other honeycomb is cut with a hive tool into a bucket, leaving only a small amount of wax on the top bar. This small amount of wax and honey residue helps to compel the bees to draw out their comb with the same alignment as before. If a comb was cross-combed or otherwise curvy, we make an effort to scrape off all of the wax residue so that a new and more aligned pattern can be established.

Combs full of honey must be handled delicately and can be lifted only with their weight directly in line with gravity, otherwise they will break and fall. Either end of the top bar may be lifted upward, but the top bar should never be rolled or tilted to the side. This is a good general rule to observe at all times when handling combs, but it becomes especially true with the added density of honey, which is by volume heavier than water. For beekeepers who are used to the rigid construct that Langstroth hives provide with their frames and foundations, this can be particularly difficult to get used to and will present itself as a distinct disadvantage in the beginning.

We use soft weeds to brush the bees from both sides of the comb and top bar and make sure the comb is bee-free before cutting it into the bucket. It helps to brush off the bees back into the hive directly above it so that they do not drop to the ground. This is much less disorienting for them and will cause less alarm and havoc for the hive, not to mention lessening the likelihood of having multiple bees climbing up the legs of the beekeeper. Keeping the lid on the honey bucket between harvesting the combs will help to prevent bees from diving in and drowning.

FIGURE 6-2. Grass being used as a bee brush. Many synthetic tools are made for this purpose, but because one or two uses makes for a sticky tool, we find that nature provides the best continuously renewable resource for this purpose.

During a good nectar flow, hives can grow rapidly, and it is best to harvest regularly, spacing the hive at the same time. Once harvest season begins, we routinely check the hives once every two to three weeks. This is often enough, as long as they are spaced appropriately at each harvest, leaving them room to grow without becoming cross-combed in our absence. Until a top-bar beekeeper is comfortable with the technique of spacing as a part of ongoing hive maintenance, it may be necessary to check the hive more often than this to establish a sense of how nectar flow and environmental factors influence hive growth patterns.

Sometimes when nectar flows shift from one predominant flowering plant to another, there will be a noticeable difference in the color of the honeys. If the honeys are not distinctly different in flavor, it may be irrelevant. However, if one honey tastes like peaches with cinnamon and is a deep orange and another tastes like clover with a hint of lemon and is light yellow, it may be to the beekeeper's advantage to harvest these honeys into separate buckets. They can then be bottled and marketed separately as unique varietal honeys. (See the Varietal Honeys section that follows.) We always bring multiple buckets in case this occurs.

Processing Honey

When we have finished harvesting, we haul the heavy buckets and the hive with honeycomb back home and process it as quickly as possible. The slow crystallization begins immediately after harvesting, so we want to bottle the honey as soon as we can. It is important to work with the honey in a room that is safe from mice, ants, and other insects, including bees. Cleaning up the floor with soapy water and then a little lavender water can keep ants out. The combs that were cut into the buckets are hand-squeezed over a mesh strainer that rests over a tank. Cells filled with pollen also are crushed, imparting their flavors and colors to the honey. The wax is left to drain for a day or two, then is removed and put outside for the bees to lick clean before it is put into the solar melter.

Once the honey is in the tank, it must be bottled as quickly as possible, again to prevent having to bottle once crystallization is under way.[1] Crystals

FIGURE 6-3. Many hands make light work. Comb-crushing draws everyone to the tank sooner or later.

in the honey make it flow more slowly from the tank spout, making the bottling a more tedious and time-consuming task. The honey also can crystallize to the point of needing to be scooped out of the tank and into the bottle—a messy and labor-intensive process. In the heat of summer, this is not as much of a concern as it is later in the fall. Once temperatures begin to cool, the honey becomes more viscous, drains more slowly from the comb, and takes longer to bottle. If the honey crystallizes in the comb while waiting in the bucket for crushing, it becomes impossible to separate the honey from the wax, and yields will be lower.

Often we are asked if crystallized combs can be heated to extract the honey. Although honey melts at a lower temperature than wax, it is difficult to apply heat evenly to the crystallized combs so that the honey will uniformly separate from the wax. Not to mention that honeybees have infused the honey with numerous enzymes that heat will denature.[2] We don't recommend applying any heat when processing honey. It is better simply to couple harvesting and processing in close succession to avoid potential problems. If, for any reason, the honey crystallizes in the hive or before it can be separated from the wax, these combs can be stored and fed to hives that need a boost going into or coming out of winter.

Because honey is a very acidic product, we bottle it only in glass jars. Honey absorbs flavors, and sensitive individuals will taste the plastic that the honey has been stored in. Although plastic honey containers can be very artful and pretty, and squeeze bears are very popular with children, raw honey is best put into a wide-mouth glass jar. If it has never been heated, it often will crystallize before it can be entirely consumed, and the customer will be grateful for the easy access. If you do opt to use plastic, it should be food-grade plastic because this is stable enough to contain the acidity of honey for a period of time before it begins to degrade. Nonetheless, we do not recommend it, not only because of the potential contamination of the honey but also because of plastic's long-term effects on the health of people and the planet. A common image that makes us cringe is one of a plastic honey bear being microwaved in order to reliquify the honey (while also degrading the plastic molecules). Yikes!

Comb Honey

When we find beautifully capped honeycombs that have been freshly built with clean, light-colored wax, we don't put them into the bucket for crushing, but rather transport them home fully intact inside a hive lined with a towel or plastic bag. Once we get them home, we lay them onto a tray and cut them with a 4 by 4 inch (10 × 10 cm) square tool that is widely available in beekeeping supply catalogs. The cut comb is then lifted with a spatula into a plastic clamshell container and put into the freezer until it is time for market. Freezing the honeycomb kills any wax moth eggs that might have been present in the honeycomb. If the honeycomb is left at room temperature, the wax moth would hatch and eat the product, so all honeycomb should be frozen before it is brought to market.

The market for fresh honeycomb is unique from the market for bottled honey. Many people seek honeycomb because they know it is truly raw and unadulterated. Honey often is mixed with other syrups before it is brought to market, and honeycomb clearly is free of this kind of tampering.

Varietal Honeys

One of the beauties of beekeeping is seeing the diverse palate of tastes, textures, and colors that flow into the hive. Honeys can range from as clear as water to as black as molasses, with tints of red, yellow, and even green.[3] Each has its own unique taste and marketability. One of the advantages of top-bar beekeeping is that each honey can be isolated and bottled separately, leading to the fun process of trying to find out the floral source of the honey and giving it a name. Some flowers bloom every year, and others are much more sporadic, yielding rare honeys.

Here's a good example of a rare honey yield. I had been keeping bees for about four years, and I thought that I knew the routine of the year pretty well. My bees typically started making honey about mid-May, and the bulk of the honey flowed in during July and August. But this particular year the bees seemed frenzied at the beginning of April. I looked into the hives and saw that they were rapidly filling and the combs were dripping with watery nectar if I tipped them sideways. I didn't notice any flowers blooming. I called a man named Ersel, who had been keeping bees for more than thirty years in my area, to ask him what he thought.

"Well, it must be a phacelia year up there in Bernalillo. Go see if the bees are carrying black pollen. Actually, it's a dark purple if you look close. The honey will be clear like water, but it will get thick and crystallize fast, so you got to extract it as soon as it quits splattering out of the combs—you can't wait until it gets 100 percent capped. That bloom will make the bees swarm crazy! Some hives will swarm three or four times! You got to give them lots of comb-building to do. Divide them once if you have to, and then give them empty frames to draw out. Lots of comb-building might take the swarm craze out of them. That flower is a little purple weed that only blooms about every fifteen to twenty years, when we get a big rain in the spring."

I checked the bees and, yes, they were bringing lots of dark purple pollen into the hives. I saw some hives with multiple swarm cells and I made a few divides. I caught nearly twenty swarms, four from my own bees and many more in town. The honey was indeed as clear as water and thick. It crystallized within a week of harvesting it. Later that year, many beekeepers complained that their hives had swarmed so much there were hardly any bees left in the hives, and that the little bit of honey they made had crystallized in the comb so that they couldn't harvest it. That's when I first got excited about starting a beekeeping association, because I realized that Ersel's knowledge should have gone out in a newsletter to as many beekeepers as possible, so they too could have been prepared for the phacelia bloom.

Marketing Raw Honey

At this time, there are no regulations in the United States to legislate the definition of raw honey, and the subject is one of regular debate among beekeepers and consumers. Honey that never has been subjected to heat of any kind is considered truly raw, and honey that has been only lightly heated or hasn't been heated past 107°F (42°C) may also be considered raw. However, because there is no legal definition of the word *raw*, even honey that has been pasteurized with high heat may be labeled as raw without any legal consequence.

Most consumers are used to the bulk liquid honey that is sold in supermarkets across the world. This honey has been heated and strained to eliminate all potential for crystallization. Often it is not 100 percent honey, but also may contain numerous fillers, such as corn syrup and other substitute sweeteners. Consumers who are used to honey as a liquid product need to be reeducated in order to buy honey that is crystallized.

Honey that never has been heated at all will, at some point, go through a process of crystallization. Although there are certain honeys that never crystallize and are coveted for this property, most all honeys will crystallize over time. Depending on the floral source and the amount of pollen and propolis particles in the honey, the resulting crystallization may render a product that is anywhere from creamy and smooth to hard and granular.

In order to educate our consumers about honey, we provide tastes of our different varietals in both their liquid and crystallized form. Because raw honey that is hand-crushed is rich with the flavor of pollen and propolis, and is still alive with enzymes, most people who taste it will be amazed at how good it is, as well as how different it is from commercial products. By providing samples and also educating our consumers about the health advantages of eating honey that has not been heated, we can easily overcome the mainstream perception of how honey *should* look and taste.

When our customers understand that they are eating a product that contains a diverse array of vitamins, minerals, and enzymes, and that is considered by many healing traditions to be a superfood, they begin to see raw honey as a valuable addition to their diets. Raw local honey often is used to combat allergies, and top-bar honey that is crushed along with the cells of pollen provides a useful form of homeopathic medicine for allergy sufferers. The homeopathic theory suggests that by ingesting small amounts of pollen from plants that cause allergies, our bodies slowly develop an immunity to these allergens. We have heard many testimonials at market from people who have effectively used our honey for this purpose.

When labeling our honey, we make sure to include that it is raw or unheated, that it is free of antibiotics and miticides, and that it is unfiltered. We also include the date and place of harvest, so that customers who are seeking honey from locations close to where they live can find what they are looking for.

Demand for raw honey is increasing as the word spreads about its healing properties. We routinely have customers who exclaim that our honey is the best that they've ever had and, although it is costly, many of our customers have our honey shipped to different parts of the country in order to keep themselves in supply.

Now that we have been selling our honey at market for many years, we have a large body of educated customers, many of whom eye liquid honey with suspicion and will ask if it has been heated. Oftentimes heated honey has a distinctly different taste and texture, and a trained palate can discern the level of caramelization that has occurred. Similarly, honey that has been exposed to hive fumigants or other chemical inputs will have lingering flavors that are noticeably out of place.

By providing raw honey to the community, beekeepers are providing a valuable service because raw honey has become a commodity that rarely is found anymore, having been replaced by much more affordable and easily produced sweeteners. Truly, honey is a gift from the bees and deserves to be enjoyed the way that nature has provided it to us, directly from the hive.

Beeswax

After the honeycombs have been crushed and the honey has been strained out of them, there will be a clump of beeswax left in the strainer. This beeswax has pollen, propolis, cocoons, and honey still clinging to it. If the combs have been thoroughly drained for forty-eight hours or more, there will be very little honey left in them. This clump of wax is put outside in a shady spot for two or three days so that the bees and wasps in the area can clean out the remainder of the honey and pollen. We do this so we don't waste any of the bees' precious resources but also because, when this residue is melted, any remaining honey will be ruined for human or honeybee consumption by excessive heat (150°F, or 65.6°C). Often beekeepers will use the honey that is left after melting wax to feed back to their bees, but studies have shown that this can cause a dysentery-like condition in the bees. It is better for them to consume this honey before the melting process, when it is still genuinely raw.

The area where the honey-wet wax is placed should be away from the beeyard and human traffic. It will become a teeming flight zone of honeybees and wasps, and in times of low nectar flow it can cause a robbing frenzy that can lead to the robbing of hives. Yellow jackets and other wasps will follow honeybees back to their hive and launch a full-scale assault on the hive if it isn't well guarded, eating all of the contents, including the honeybees and brood nest.

(Note: If a hive is being attacked by yellow jackets or other wasps, it is necessary to make sure there is only one entrance to the hive and modify it so that only one or two bees can access the hive at a time. This modification can be done by plugging a large portion of the entrance with beeswax, tape, or wood. If propolis is available, this is the best substance for this purpose. Reducing the size of the entrance and exit to the hive gives the colony a chance to guard itself more effectively against the attack.)

Often we will transport the wax to and from the cleaning area early in the morning or after dark, so that we aren't handling the wax container when it is active with bees and wasps. Beekeepers who are uncomfortable with this model of cleaning the wax can instead place chunks of wet wax into the back of a hive, where it can be cleaned by the hive in privacy. Then it can be removed after a day or two for melting. This avoids feeding the local wasps and obviates the need for an area that would inevitably become a honeybee traffic zone.

The beeswax becomes valuable once it has been melted and cleaned of debris. The sun provides ample energy to fuel this process, and there are numerous solar-melter designs available online. They can be tailored to fit a beekeeper's individual needs in terms of size and capacity. The simplest melter is made from a stainless steel or enamel container covered with a sheet of glass. The container should not be made of galvanized metal or iron because the exudates from these metals will contaminate and discolor the beeswax. Plastic tubs are unsuitable because they warp and degrade in the intense heat. Aluminum is also a poor choice because it is slightly soluble in contact with the acidity of honey. We use an enamel tub and a sliding glass door that we bought used from our local Habitat for Humanity Restore.

The container is tipped slightly toward the sun, and a quart or so of water is placed in it. The wax debris is placed on the high side of the container in the morning before the sun is high. The sheet of glass is secured over the container so that honeybees cannot get in and the heat cannot escape. The heat of the sunshine liquifies the beeswax and any remaining honey in the debris and they trickle down into the water, where the wax will float and the honey will dissolve. The next morning, the beeswax will be a solid sheet floating on top of the water and can be lifted off and collected into a bucket. The cocoons and remaining debris will still be at the top of the container and can be removed and composted. There is a time in the morning when the sun has heated the cocoons so that they can be scraped out easily, but the wax has not yet re-liquified. This is the easiest time to do this job. Then a new batch can be put in.

The wax collected out of the solar melter is not yet ready for candle-making. There is still a level of debris in unstrained beeswax that will plug up the wick in a candle and make the candle burn dimly or with a great deal of dripping. We usually store the wax that comes directly from the solar melter until all of the wax from the season has been processed in this way. Then we move on to the second stage of filtration. Eventually the unfiltered beeswax is melted in a double boiler so that it can be filtered.

Wax should never be heated over a direct flame. This can cause the wax to scorch and is a potential fire hazard. Beeswax that has been scorched will blacken and the delicious honey smell will disappear, to be replaced by a burnt smell. For small amounts of wax, a double boiler placed on a stovetop works just fine, but since we have a large quantity of wax, we float 5-gallon (19-L) buckets in a large stainless steel tub filled with water and outfitted with a submersible heater. We wrap the tub with insulation in order to preserve the heat while we filter the wax and use the same buckets to dip candles in. Because we are dealing with large amounts of wax, we want to process as much as we can at a time in order to save energy.

We have found it best to pour a little water into the 5-gallon buckets of wax. The water absorbs impurities and makes it easier to remove the wax when it has hardened. The wax will melt at 150°F (65.6°C). During the melting process, we leave the wax undisturbed so that the impurities either will settle at the bottom in the water layer or will float to the top and can be skimmed off. When the melting process is finished, there will be a top layer of wax with some debris, an interface layer of wax and water with lots of debris, and a layer of dirty water. The top layer is decanted carefully into a bucket with a paint strainer secured over it. Other materials, such as clean t-shirt material, also can be used for straining, but paint strainers are inexpensive and readily available for this purpose. It is important to pour only the relatively clean layer of wax and to leave the debris-filled interface layer in the bucket with the water. (The debris in the interface layer will plug up the strainer quickly and prevent any more wax from being processed.) Once it has cooled, the wax left in the interface layer is put back into the solar melter to be reprocessed.

If a beekeeper skips the solar-melting process and puts the crushed combs directly into the double boiler, the result would be one big interface layer from which little or no clean wax can be strained. It is very important to go through the solar-melting process to remove the cocoons before placing the wax into a double boiler.

As the honey-making season progresses, there may be color changes in the beeswax. Just as different nectars change the flavor and color of honey, the nectar and pollen flows render different colors and scents of wax. Beeswax can vary in color from very dark brown to reddish orange to lemon yellow. When we have divergent colors of beeswax, we make an effort to keep them separated, just as we would with different types of honey. If very dark-colored beeswax is undesirable to the beekeeper, it can be bleached with hydrogen peroxide. Much commercial beeswax has been bleached, because it is commonly derived from beeswax combs that are discarded once they have turned black and are no longer suitable for honey production.

Top-bar hives provide a lot of beeswax, which can greatly add to a beekeeper's income. If the wax has been produced in hives that have not been treated with miticides, it becomes exceptionally valuable because it is free of miticidal toxins. This is an important quality for wax that is used to make cosmetic products because contaminants in the wax can be absorbed by the skin. We market our beeswax as *unbleached and miticide-free.*

Once a good quantity of pure clean beeswax is collected, either we can sell it by the pound or we can make things that can be sold directly to customers for a retail price. Once our busy summer harvest season is over, we spend the fall and winter making candles, beeswax soaps, herbal-infused healing salves and lip balms, and beeswax ornaments. These products contribute to our income and take care of our customers' desire to buy local, healthy products from chemical-free hives. The book *Super Formulas, Arts, and Crafts: How to Make More Than 360 Useful Products That Contain Honey and Beeswax,* written by Elaine White, has many ideas for beeswax products.

Beeswax candles are a very popular item. They smell wonderful as they burn, and they burn longer than paraffin, which is a refined petroleum product. Although a candle is a simple device with only two ingredients, there are some things to keep in mind in order to make a candle that will burn well. The wick comes in many sizes, and its function is to soak hot liquid wax up out of a pool of wax and mix it with air so that the wax will burn evenly. The size of the wick determines the size of the flame and the amount of heat produced. A very large wick in a small candle will produce an intense flame that burns quickly, like a torch. A wick that is too small for a candle's width provides a flame that's insufficient to burn off the wax it melts and it will drown itself. The level of wax refinement also makes a difference because any impurities (in our case, bits of propolis, dust, or hive debris) float into the wick and plug it up, causing irregular burning and drips.

When making candles with beeswax, the wick needs to be larger than a wick rated for refined paraffin. When dipping tapers with a ¾-inch (1.9 cm) base, the #2 square braided wick is a good match. For molded candles larger in diameter than 1 inch (2.5 cm), we use a #5 square braided wick. For very large candles, we twist or braid several wicks together. When working with experimental shapes and sizes, it is important to burn one of the candles to see whether it burns too quickly, drowns itself, or is just right. If the candle burns slowly and develops a pool of liquid wax, a clump of ash may form at the top of the wick after a certain period of time. This ash is falling off the wick into the liquid wax and then getting pulled back up as the wax burns. The ash does not burn and eventually plugs up the wick. When this happens, it is best to blow out the candle and cut the clump of ash off of the wick. If the ash falls into the liquid wax, it can be poured off.

Beeswax is also an important ingredient in many healing salves and balms. It is a natural emollient that seals and nourishes dry skin. If beeswax is melted into oil, the wax will stiffen the oil base. A ratio of approximately five parts oil to one part wax generally will make a product that will be soft enough to rub into skin and yet solid enough that it doesn't run all over. There are many healing plant oils, each with unique chemical and physical properties that may blend and react in different ways with beeswax, so it is helpful to test various formulas to create the desired consistency.

Propolis

Propolis is an amazing substance and is collected by bees specifically to disinfect and seal their hive body. It is a highly antimicrobial resin that trees exude to protect their buds from infection. Honeybees collect it by scraping at the surface of the plants with their mandibles and transport it in the pollen sacs on their legs to the hive, where it will be distributed where it is needed the most.

Propolis, or *bee glue*, as some beekeepers call it, is gathered throughout the year and can be moved around the hive for various purposes. Most often beekeepers will find it between and under top bars, gluing them together and adhering them to the hive body. On closer inspection, propolis also can be found on the inside of the beeswax cells and as a coating for the interior surface of the hive, where it has been applied lightly by the bees as a disinfectant.

A hive tool comes in handy to dislodge a tight propolis seal in the springtime or after a cool night in the fall. However, once a hive has been sealed for winter and the honey harvest is over, it is best to leave this propolis

FIGURE 6-4. A honeybee carrying the distinctive red propolis from the trees around our home.

seal undisturbed because it provides the bees with the conditions they are most comfortable with. Any disturbance may mean that some workers must break cluster to repair the damage or air leak that may have been left when the seal was broken.

Propolis is known for its powerful antibacterial properties and can be found in many health products, from toothpaste to tinctures.[4] The bees use it similarly to keep their hive from infection. If a mouse happens to get into a hive, the bees will sting it to death and completely encase it in propolis. The propolis effectively mummifies the mouse, keeping the decaying process from moving forward and protecting the bees from bacterial infection.

Honeybees instinctively will want to propolize closed any openings that are not useful to them, and propolis traps are designed to capitalize on this instinct. They are made of a plastic grid that has holes too small for a bee to move through, but large enough that the bees will want to have them closed for comfort. Thus, the bees will fill the holes with propolis, and when the trap is completely coated with propolis, it can be removed or harvested.

When it is warm, propolis is extremely sticky and can stain and penetrate most materials, so the propolis is not removed from the trap while it is warm. It is placed into a refrigerator or freezer where it will solidify. Then the trap can be twisted just like an ice cube tray, and the propolis will crack off in brittle chunks that can then be processed into powder and sold by weight.

Springtime is the best time to collect propolis because the hive entrances and cracks will be loaded with it and it is easy to scrape away large chunks with a hive tool. We carry an empty yogurt container with us in the springtime expressly for the purpose of storing propolis. We then sell all of our propolis to an herbalist friend of ours, who will take whatever we can provide him with.

Propolis can be eaten, but it has an undesirable texture and can stain the teeth. It also can be very spicy or pungent in flavor, depending on which plant it has been derived from. The color varies from green, yellow, or brown to bright orange or red. In our location here in the mountains of northern New Mexico, the red propolis is hot and can burn the mouth with its heat.

The tendency to gather propolis varies from one race of bee to another, and although some bees are heavy propolizers, others are not. The darker races of bees from colder climates are said to be better propolis gatherers.

Royal Jelly

Another beehive product customers often request is royal jelly. It is the special food that queen larvae are fed that causes them to develop into queens. It is prized as a superfood because it contains a rich supply of B vitamins and many life-enhancing enzymes and minerals. Traditionally, it has been used as a tonic for vitality and fertility, particularly in China, where a race of bees has been bred specifically to produce copious amounts of royal jelly.

Honeybees make royal jelly at a great cost to the hive. Queen-raising must be initiated and seventy-two hours later each queen larvae is pulled out of its cell and the cell is suctioned of its royal jelly. This extremely labor-intensive project yields very small amounts of product. Even the bees that have been specifically bred for this purpose must be rested frequently in order to recover their strength to begin the process again. When working through a hive, we occasionally break open a queen cell and taste the royal jelly, but we do not produce it or sell it.

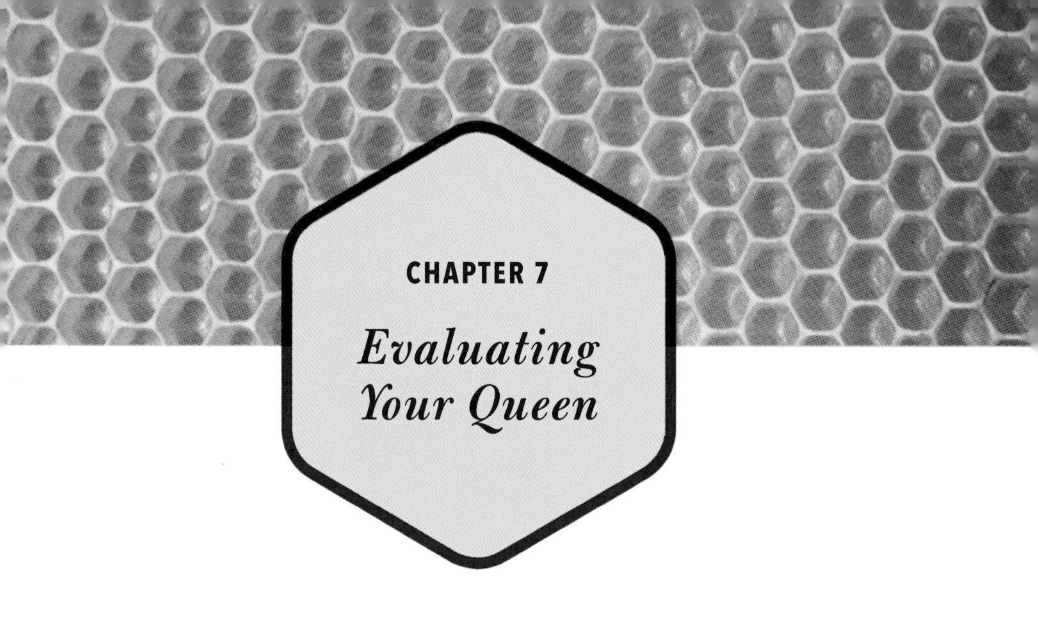

CHAPTER 7

Evaluating Your Queen

Often when we begin beekeeping, we are grateful simply to be able to find and identify our queen, and it is enough to know that she is there. However, as we develop as beekeepers, it becomes increasingly important to be able to evaluate a queen's strength because it may be a deciding factor in whether a hive makes it through winter or is able to be divided in the spring. It also determines whether or not a hive makes honey.

A good strong queen will be large in size and, after she has mated, she will have a firm, turgid abdomen full of eggs. Occasionally one may see an unmated queen in a hive in which she has just hatched. Her abdomen will be somewhat smaller and should not be seen as an indication of her future strength. Only after she has mated and is laying eggs can she be evaluated accurately. Once she is mated and laying, however, she definitely should look substantially larger than the workers in the hive. The larger the queen, the more likely that she will lay out well and for a long period of time. Although small queens may perform well for a period of time, they are generally more short-lived in their egg-laying than larger queens.

A good queen will lay a very tight pattern of eggs in the comb, and there will be only one egg in the center of each cell. Occasionally a queen that has just started laying will lay two eggs in a cell, but this is unusual and often doesn't bode well for the future. A strong queen will lay a very even spiral of eggs in a comb and will be extremely prolific once she has started laying. In a divide or a nucleus hive in which there are very few bees to care for the eggs or a small number of combs in which to lay eggs, she will be harder to evaluate because she cannot show her true strength until the number of workers builds to accommodate the full strength of her egg-laying.

A dearth of nectar and pollen also can affect how well a queen performs. If the hive is stressed, the queen will be stressed and often what looks like a poor queen can evolve into a strong queen once the nectar and pollen begin to flow again.

If a queen has a spotty pattern and there are eggs and multiple ages of larvae on any given comb, the queen is weak and would best be replaced before winter. It is important that high standards are used in evaluating queens because, over time, the hive's strength will diminish if she is weak. As the summer passes, valuable time and nectar flow that the hive might have been building on will be lost. Remember that the queen is only one member of the hive, and when she is replaced, the whole organism potentially has been saved.

FIGURE 7-1. An example of a failing queen. Not only does she have an extremely spotty egg-laying pattern, but there are also signs of disease in some of the larvae, as evidenced by an off color and slight slumping in the cell. Often disease will manifest when a queen is ready to retire and has grown weak. Providing the hive with a new and healthy queen often eliminates any sign of disease.

Requeening

When we need to replace a queen, we locate her and carefully squeeze her head and abdomen so that she is killed quickly. We then place her on the bottom of the hive so that the bees can find her there and begin their lament. We do this rather than casually discarding her, out of respect for whatever process the honeybees may have for acknowledging her death. Within a matter of hours, the whole hive will be informed of her absence, and they will begin to mobilize an effort to raise a new queen. If introducing a caged queen to the hive, it can be helpful to wait until the day after the old queen has been killed, giving the members of the hive time to adjust to their new condition. This may help with their acceptance of the new queen. If their old queen was poor or ill, they readily will adopt a healthy, newly mated queen in her place.

It can be psychologically difficult to kill a queen, especially one that has been a strong and vibrant member of the community. In a natural environment, the bees would take care of this themselves, *balling* or stinging to death a queen that is no longer serving the well-being of the hive. Unfortunately, we are seeing a decline in the natural instincts of the bees to requeen themselves, and many hives will follow an ailing queen to their own demise. This may be due to overdomestication or artificial breeding techniques and would be worth some further study.

If a beekeeper is really averse to killing the queen but wants to save the hive, the queen can be taken out in a small divide and left to diminish slowly in peace. The original hive then can be saved by having retained most of its numbers and being given a strong, healthy, newly mated queen.

Honeybees are able to detect the scent of bees from another colony, and they often will repel or kill foreign bees. When we requeen a hive, we are asking the colony to accept a foreign queen. There are factors that influence queen acceptance, and taking these factors into consideration when requeening can lead to a much greater rate of success.

- Queen introduction is easier when flowers are blooming. Worker bees are more accepting of foreign bees when they are busy making honey. Timing the introduction of a new queen with a nectar flow is definitely helpful.

- A hive will not accept a foreign queen if it has its own queen or developing queen larvae, so both of them must be removed. Developing queen larvae supply queen pheromones to the hive and will reduce the acceptance of the new queen somewhat. The larvae can be flicked out of the queen cell with a twig, providing a rare opportunity to eat the royal jelly.
- Queen acceptance is also greater when there are worker brood pheromones in a colony. (Drone brood does not have the same effect.) If there is no brood in the hive, it is extremely helpful to acquire a comb of brood from another hive (without the bees) and the queen cage can be secured to it.
- One of the easiest ways to place a new queen is to put her into a divide. Divides mainly are populated by newly hatched nurse bees who are much less discriminating than established foragers.

When the first Russian honeybees were brought into the Americas for their mite-resistant traits, there was a low rate of acceptance of the Russian queens by the American bees. It was hypothesized that the Russian honeybees were genetically so different that it was harder for the American bees to take them in. Once extra precautions were taken by putting the Russian queens into divides with young bees and larvae, the acceptance rate went up. Now Russians are part of the American honeybee genetic pool, providing us with valuable mite-resistance and cold-hardiness traits.

When placing a new queen into a hive, the cage holding the queen should be pressed into honeycomb just above the brood. This ensures that the nurse bees will be around her while they tend the brood. The cage screen allows them to feed her but not sting her. The candy-filled doorway should point slightly up and to one side. The plug or cap protecting the candy from the hive bees is removed so that the bees can begin eating the candy. If the cage has no candy, the door to the cage is kept closed. The new queen should be in the cage for three to four days so that the hive can acclimate to her new scent. The candied doorway is designed to last just about that long. When she is ready to be released, the cage may be placed on the bottom of the hive and gently opened so that the queen can enter the hive without raising an alarm pheromone.

Requeening Aggressive Hives

Sometimes a hive that was once gentle becomes aggressive. This can happen because the hive has been allowed to swarm freely and raise its own queen, who then went out and mated with a drone with aggressive tendencies, which then transferred to her offspring. When a hive begins to show aggressive traits, it is important to requeen the hive as quickly as possible. Letting hives remain aggressive becomes a liability to the local gene pool and potentially to neighbors and friends who find themselves in the bees' flight path.

I once knew a woman who had hives in the city of Albuquerque. She'd kept bees for years and never had any problems. One day she called me out of the blue to tell me that one of her hives had become so aggressive that she was afraid to work them anymore. It was her strongest and most productive hive, and she was loath to get rid of it, but they were stinging her through her clothes and she was beginning to feel as though she "had a monster" in her backyard.

I recommended to her that she move the hive to a more remote location in order to go through the process of requeening them, but she was reluctant to let them go. The next time I heard from her, her neighbors were threatening to sue her because they'd had a party in their backyard, which was adjacent to hers, and many people were stung. In a panic, she went out at night, doused the hive with kerosene and covered it with a black plastic bag.

It is best to deal with aggressive bees before the problem becomes this accelerated. As soon as signs of aggressive behavior are detected, the hive must be requeened. It is important to be able to distinguish when a hive is having a hard time and is being cranky due to a lack of nectar flow, and when it is truly becoming mean. If it is consistently aggressive, no matter how beautiful the weather or environmental conditions, it needs a new mother.

The best way to handle an aggressive hive is to divide and conquer. The hive initially is moved to a new location in the beeyard and an empty box is put in the old location along with a few combs of brood from which the bees have been brushed. The older, more established foragers are the most aggressive bees in a hive, and they will be the ones who will go back to their original location to inhabit the few combs in the new box. The absence of the older bees will make it easier to go through the hive to find the queen. If working through the hive is still untenable, the hive should be divided quickly into two or more boxes simply by lifting four or five combs at a time into a new box. The more divides are made, the more docile the bees will become.

After the divides are made, they can be left alone for four or five days, at which time it will be easy to find the queen because she will be in the only hive with eggs. If the old hive has been adequately divided, there will be only four or five combs to search through in order to find her, and then she can be killed and a new, gentle queen can be introduced. The hive then can be recombined.

It is very important when recombining the hive that each comb be thoroughly searched for impregnated queen cells. Letting an aggressive hive raise its own queen is ill advised because it is likely to propagate the very genetic traits that need to be eliminated.

Robbing

There are times when we must harvest honey that has accumulated during a heavy nectar flow that has since ceased and been replaced by a dearth of nectar coming into the hive. This makes the honeybees very upset. Although we always leave ample honey for the bees, if there is no fresh nectar coming in, they will covet every last drop of honey that they have stored away. This makes them *robby*. When bees are robby, they will dive into any open hive and try to rob their neighbors' honey. Then they begin to fight and sting wildly. This is the most difficult condition under which a beekeeper sometimes must work.

At times like this, it is necessary to move very efficiently, making sure to keep the lid on the harvest bucket and to leave hives open only as long as necessary. It is important to take every precaution to keep from spilling any honey, especially around the hive and bucket. Setting the smoker next to the bucket so that the smoke continues to billow around it while the honey is being cut can help a little.

When there are a large number of hives in a beeyard and the bees are robby, it can be extremely difficult to get all the work done. For times like these, we use a mesh tent, which we drape over ourselves, the hive we are working on, and the harvest bucket. Even using this measure, the hive being worked on can be extremely cranky, stings will result, and killing bees while closing up the hive is practically unavoidable. Often when we know that we will be facing this kind of environment, we will go to work on the hives much earlier than usual, even in the predawn light of day. If it is cool enough, the bees still will be withdrawn from the honey at the back of the hive and the combs can be taken quickly before the bees are ready to be flying. Going in early can help the beekeeper to get through a good number of hives before the robbing frenzy begins and conditions become unbearable.

It is important for a beekeeper to be able to recognize the conditions that create a robbing frenzy. Often new beekeepers will think that their hive has become Africanized or mean when in fact there is a very dry period going on and their hive actually would be a very gentle hive under more lush conditions. Bees are as susceptible to bad moods as humans when they face the possibility of starvation.

Intercaste Queens

When a hive gives birth to an intercaste queen, it can be very difficult to diagnose. She is hard to see and will lay eggs that look relatively normal until they reach maturity, at which time they will all be drones. If she can be spotted, her abdomen will be queenlike but smaller and slightly triangular in shape (See Figure 7-2). She is the product of an egg that was reared as a worker bee before it was changed into a queen. She did not receive enough royal jelly to develop her ovaries completely and will never be a viable queen. In order to correct a hive with an intercaste queen, she must be found and removed. Only then will the hive have a good chance of switching allegiance to a new queen.

Drone-Laying Queens

Occasionally a hive will give birth to a queen who looks normal, but all she lays are drones. This is usually due to her inability to fertilize her eggs.

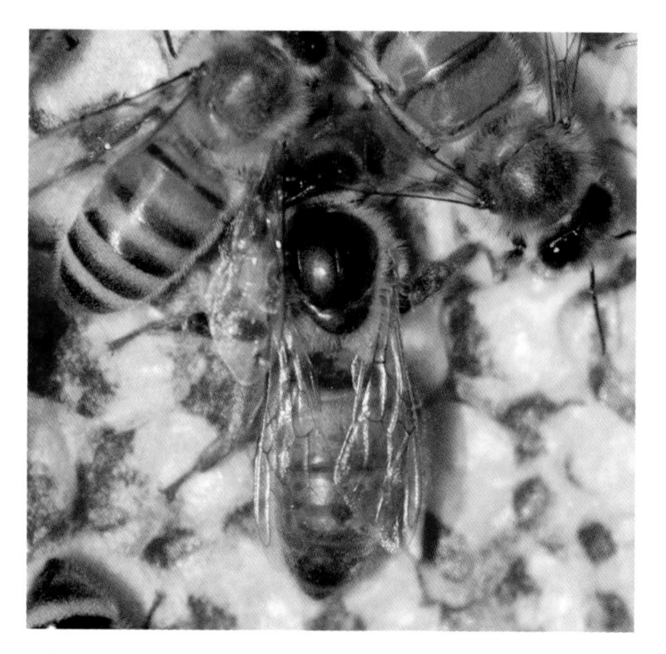

FIGURE 7-2. An intercaste queen. Her abdomen is smaller and more triangular in shape than a fully developed queen.

A hive also can develop a drone-laying queen over time if a queen runs out of semen and no longer can provide fertilization. If drones are being hatched out of worker cells, the hive either has laying workers or a queen who is able to lay only drones. The only way to remedy this problem is to find her and replace her with a viable queen.

Laying Workers

If a hive is unsuccessful in raising a new queen and it remains without brood or a queen, eventually it will develop laying workers. Due to the absence of a queen's powerful pheromones, female worker bees will begin to develop their ovaries and will lay an abundance of unfertilized eggs. This is the hive's last effort at spreading their genetics into the world. The hive eventually will die out as the worker bees perish and the hive's resources diminish.

FIGURE 7-3. Laying workers will fill cells with an abundance of eggs, in contrast to a healthy queen, who will deliberately lay one egg in each cell.

It is fairly easy to recognize a laying-worker hive because they will build only drone cells and the comb will have a very bumpy, irregular look to it. There will be a nonuniform pattern of eggs scattered here and there, and often there is an abundance of eggs in each cell.

It is difficult to save a laying-worker hive. Simply giving them a caged queen has a very low rate of success. Once they start laying, they aren't particularly interested in accepting a new queen into the hive. If the workers have only just begun to lay, giving them a comb of eggs can sometimes compel them to raise a queen on their own. This is only moderately successful but certainly worth a try.

Another approach is to shake out all of the bees about 50 feet (15 m) from the hive and then put a caged queen into the hive with the bee-free combs. The idea behind this method is that the workers who are laying eggs are too heavy to fly back home and will be unable to make their way back by crawling. The workers who haven't yet developed their ovaries will fly home and are more likely to accept the new queen who is waiting there. We have found that this method is only moderately successful as well.

It is also possible to combine a laying-worker hive with a strong hive. By putting the laying workers at the back of the strong hive and erecting a paper wall between the two hives, the colonies slowly will combine with one another and the pheromones of the queen in the strong hive will dominate, suppressing the laying workers from continuing to lay eggs. Many of the flying bees from the drone-laying hive will go back to their original location, so it is best to take their original hive completely away so they will be forced to seek acceptance into a hive that is close-by.

The method we use most often is simply to shake out all of the bees and take away their box and combs. The combs are added to other hives in the yard and the bees thus are forced to join the other hives in the beeyard. Because they come as intruders, they are humbled into acceptance through an initiation process whereby the hair around their thoraxes is chewed away to mark them as foreigners. Although this is the method we most recommend, it is not always an option if there aren't other hives available for them to join.

Turning a hive from a laying-worker hive into a viable queen-right hive is a challenge, so it is better to do regular prevention. This can be done by checking for the queen each time we enter a hive, not necessarily by finding her, but rather by checking the brood pattern for eggs. If only larvae or capped brood can be found, something has happened to her and it will be necessary to provide the hive with a queen or eggs to raise a queen from immediately. Prevention is the best cure when it comes to laying workers.

CHAPTER 8
Problem-Solving

As beekeepers who advocate strongly against the use of antibiotics, miticides, or other chemical applications within the hive, we need to have remedies for hives with problems. Our most common remedy is simply to requeen. Requeening a hive can solve problems that range from weak egg-laying to aggressive behaviors, as well as susceptibility to many diseases, mites, and parasites. By doing this, we are in a constant process of breeding stronger and more disease-resistant bees. We believe that by treating bees with medicines or other inputs we are simply putting off what is ultimately a genetic weakness that will reoccur.

In order to know when to requeen a hive, it is important to be able to recognize the signs of various diseases.

Chalkbrood

Chalkbrood is caused by a fungus that preys on the larvae in a hive, turning them into moldy-looking, chalky mummies. Recent research has shown that chalkbrood can be caused by the application of fungicides in areas where bees are foraging for pollen. The fungicide in the pollen inhibits the naturally occurring fungi that bees use to preserve their food and creates adverse conditions within the brood nest, providing a pathway for disease.

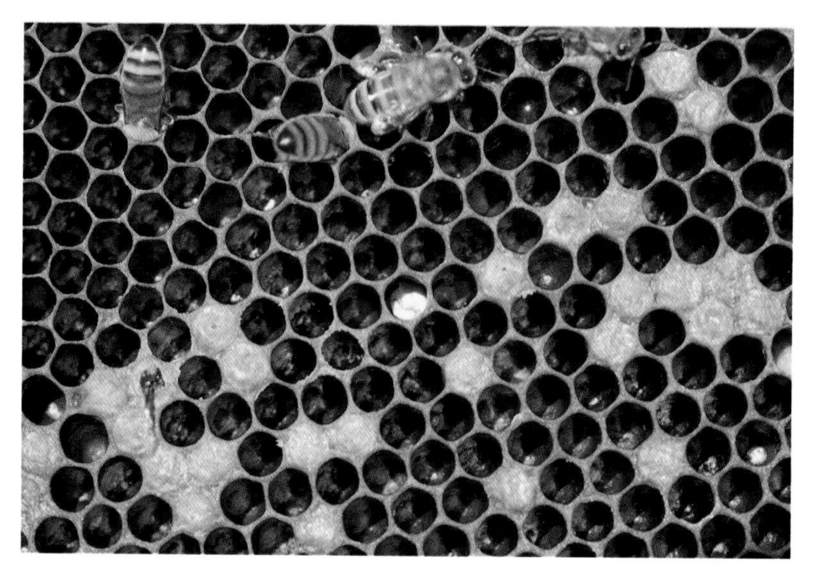

FIGURE 8-1. A hive showing signs of chalkbrood. In the center is a larva that has turned chalky. Hives with mild symptoms sometimes will outgrow chalkbrood, especially if there is a good nectar flow going on, but more often they require requeening.

We had direct experience with this phenomenon when we took our hives to pollinate almonds in California. Although we were careful to choose an organic farm to pollinate, the environment in the Central Valley of California is so contaminated with fungicides and other chemical applications that our bees showed multiple signs of illness. Fungicides are used heavily on almonds, and our hives became largely infected with chalkbrood. By ceasing our pollination of almonds and doing vigorous requeening, we cleared up this issue in a period of approximately two years.

Although bees can be bred for chalkbrood resistance, it is important to remember that requeening is not always going to solve the problem in cases of chemical contamination because combs, stored pollen, and honey also may harbor continuing pathogens.[1] Requeening combined with the rotation of all old combs out of the hive is the best way to eliminate the issue completely, along with removing the hives from the source of the contamination. If there is an extreme case of chalkbrood in a hive, it may help to take away as much pollen and honey as possible before requeening, provided that there is a healthy nectar flow coming into the hive.

There was a time when chalkbrood was not much of an issue, and it was rare to see it at all. Now it can be found more readily in hives and should be considered a significant problem to be eliminated. If requeening doesn't help and it continues to reoccur, it might be necessary to move the bees to a new site because there may be an unidentified regular contaminant in the surrounding area.

Foulbrood

Foulbrood is the disease that beekeepers fear the most, and it is the reason that beekeepers administer antibiotics to their hives on a regular basis. When foulbrood (European or American) affects a hive, the brood begin to die. They turn a brownish yellow color and slump in their cells. Their bodies become mucuslike and they smell very foul. As foulbrood progresses in a hive, more and more larvae die until the hive is destroyed. Whole beeyards can be lost to this disease if precautions are not taken immediately to prevent it.

Before antibiotics were widely used in beehives, the recommended treatment for hives with foulbrood was to burn them. The disease is recognized as highly communicable, and state and federal agencies have used hive burning to prevent epidemic outbreaks. Many hives have been destroyed in this way. When antibiotics became available, beekeepers started to use them prophylactically in order to prevent foulbrood and other bacterial diseases. Many beekeepers administer antibiotics to their hives every year, regardless of the health of the hive when the medicine is administered.

There are bees that are susceptible to foulbrood and there are bees that are resistant to this disease.[2] Selecting for bees that show resistance to foulbrood is the main form of treatment that we recommend. While keeping a beekeeper in production, the prophylactic use of antibiotics perpetuates strains of bees that are susceptible to this disease. Ultimately, the bees are not being saved by this strategy and the overall genetic pool is being weakened.

Antibiotics were not designed to be used on an ongoing basis or as preventative medicine, and their continued use helps to breed strains of bacteria that are resistant to them. This creates a vicious cycle in which beekeepers and pharmaceutical companies are wed to one another in an ongoing battle against nature. Furthermore, the regular use of antibiotics jeopardizes the delicate internal balance in the honeybees' intestinal tract, leaving them more susceptible to harmful viruses and bacterial infections. Using antibiotics as a routine part of beekeeping can be a harmful dependency, both for the beekeeper and the bees, and it weakens the overall immune system of the colony.

Years ago, when I was honeybee inspector for the state of New Mexico, I made a visit to a man named Dr. Lyle in order to inspect his bees. He was in his eighties at the time and looked at me with some suspicion as I arrived at his home. He obviously didn't like the idea of the state sending a young guy to tell him how to take care of his bees. He immediately began to quiz me about my motives and what I was looking for in his hives. At first I felt irritated because I just wanted to look through his bees, file my report, and be on my way because I had many more inspections to do that day. But then it became clear that he had something to teach me.

He asked whether I was looking for foulbrood. When I said yes, he asked me what I would do if I found it. I told him that I felt that beekeepers were using too many antibiotics in their hives and that I recommended requeening with disease-resistant queens and getting rid of all of the old black combs, which usually got rid of the problem. His mood brightened when he heard that, and he told me that he was a retired doctor and that he'd done some research on the bacteria that reportedly caused foulbrood. He said, "Son, I looked up those bacteria and those bacteria, are not pathogenic bacteria; they are putrefactive bacteria. I don't believe they cause the disease; they just decay the dead larvae."

I asked him what he thought caused the disease and he responded with, "The old black combs in the hive, particularly in the brood nest, are like us trying to stay healthy in a house full of sewage. Those old combs get thick and heavy, full of molds and bacteria. What you are calling disease-resistant bees are normal, old-fashioned, decent bees. What we have around here these days are bees with a genetic deficiency that makes their brood die. If you ever see any foulbrood in your own bees, I want you to try something. The books all say that if you find a hive sick with foulbrood, you should sterilize your hands and hive tool before you open the next hive so you don't spread the disease. I want you to try to spread the disease! Take combs full of sick larvae out of the sick hive, brush most of the bees off, and put the whole stinking comb in a healthy hive. Watch what happens to the healthy hive. Keep trying if you want. I bet you won't be able to spread the disease, providing you don't fill up the hive with old black combs."

A week later, I did find American foulbrood in one of my hives. I decided to try Dr. Lyle's experiment. I shook most of the bees off and gave the comb full of dead, decaying larvae to a small, queen-mating nucleus hive. I chose this small hive because I didn't want to infect a big, honey-producing hive with the disease. I also thought that the small hive would be more rapidly impacted by the sick comb. It had a total of only four combs, and now one of them was filled with foulbrood-infected larvae.

One week later, I looked inside and there were no dead larvae or any other indication of foulbrood on the healthy combs, and the foulbrood comb had very few sick larvae left. I thought that perhaps there might be an incubation period and that I hadn't waited long enough. When I checked again in another week, the hive seemed disease-free. I gave it another infected comb and watched it all summer. It never got sick. I came to the conclusion that whether the larvae die of the bacteria or a genetic deficiency didn't matter. There are disease-resistant bees and when they receive a sick comb in their midst, they simply clean it out. By culling old black comb from my hives and requeening any hives that show signs of disease, I now rarely see any signs of foulbrood and have not used antibiotics for more than thirty years.

Unfortunately, not enough selective breeding is being done at this time, and many bees are being treated with medications rather than being bred for genetic resistance. This needs to change, and beekeepers not only ought to seek out foulbrood-resistant bees but also be willing to work on breeding their own bees for resistance. Unless more beekeepers and honeybee researchers actively get involved in this process, honeybee populations will continue to be at risk from this disease.

Tracheal Mites

Tracheal mites infect the trachea, or breathing tubes, of honeybees. When bees are affected with tracheal mites, their wings will be extended out from the body because they are unable to fold them against their abdomens. They act jittery and visibly will have trouble moving around. They are slowly suffocating and cannot fly, so they will be seen wandering about the hive and on the ground in front of the hive until they die.

The conventional treatment for this condition is to place menthol crystals within the hive. When the bees breathe in the fumes from the menthol, the mites are killed and the hive is relieved of the problem.

Many bees are resistant to tracheal mites,[3] and selecting for this trait is our recommendation. Again, if we see tracheal mite symptoms in a hive, requeening is our preferred remedy. Tracheal mites must be positively detected by dissecting worker bees, dissolving the fat around their internal tracheas, and inspecting the tracheas under a microscope.

Varroa Mites

Varroa mites came to the United States in the 1980s. Many beekeepers were wiped out completely by them, and the beekeeping community struggled to find a solution. Hardly any bees had the ability to resist the invasion, and most hives died within a year of being infected.

Varroa mites attach themselves to honeybees and suck their blood. The pregnant female mite enters the cell of a pupating larva and lays six to eight eggs. The offspring hatch, breed, suck pupal blood, and then three or four pregnant female mites emerge with the bee when it hatches and search for other larvae to infest. Their population increases exponentially, eventually causing the total collapse of the hive.

These mites also transmit a virus called cloudy wing virus, which shrivels up the wings of the infected bees. If a hive has varroa mites, they will appear as tiny red dots, often attached to the head or thorax of a bee. This is the only overt symptom of their presence unless the hive also is infected by the cloudy wing virus. Varroa mites prefer drone larvae to worker larvae for breeding, so tearing the cap off capped drone brood and searching the larval cell for mites can indicate how badly the hive is infected.

Before the varroa mite arrived in New Mexico, I had bred bees for foulbrood resistance and for tracheal mite resistance. In retrospect, this was fairly easy to do because the gene pool offered a relatively high rate of resistance to these problems. The resistance traits were easy to find by simply withholding the treatments and selecting for survivors. Varroa mite proved to be a much more difficult problem because the bees showed little or no established resistance.

In 1993 I began to see mites in the hives of friends. Then a neighbor called and said, "My bees have these funny little red things crawling all over them." Nearly every bee in her hive had at least one mite, and many had three to six mites on their bodies. They also were showing symptoms of cloudy wing virus. I knew my bees were next. By the end of that year, most of my hives had mites and about half of them died over the winter. That was a hard blow, but I had been through something similar when I discontinued antibiotics for foulbrood. In 1994 I lost only a third of my hives and I thought I was making significant progress. Then in 1995, 98 out of my 100 hives died from mite infestation and the two remaining hives were full of mites.

I tried experimenting with formic acid because it was organically approved for beekeepers in Germany. Although it was effective in killing mites, it was difficult to work with. Its dispersion rate was temperature-dependent so it volatilized too quickly in our hot New Mexican summers and would not disperse at all on cold days. It was also hazardous to use; after some splashed onto my eye and burned me, I stopped using it. I simply wasn't happy using a chemical solution for mites, no matter how organic it was.

Another treatment method that was being used was spritzing the bees with oil in an effort to suffocate the varroa mites. A careful proportion of oil needed to be used because too much would suffocate the bees as well. This was effective in staving off the mites but didn't solve the problem because it required continuous application. Dusting the bees with powdered sugar was another nontoxic method that provided temporary relief from infestations because the sugar caused the mites to fall off of the bees, lessening their impact on the hive. Although these methods were useful, they didn't provide a sustainable solution to the mite problem.

Around that time, I read about an Italian beekeeper who was keeping track of feral beehives in Italy. He wrote about how he had lamented the loss of the wild Italian bee to varroa mite predation after it had first arrived in Italy. Eight years later, these wild bees began to return to the various sites he was checking on, and within ten years he found the wild colonies had returned to their pre-varroa population levels. I had at least a small modicum of hope that we would see some resistance occur in that same time period, but in the meantime I was struggling to keep my hives alive.

In 1998 I attended an American Beekeeping Federation meeting with the hope of finding anyone who could help us with developing varroa mite resistance. Most industry and science leaders seemed to repeat Dr. Tibor Szabo's often-printed assertion that establishing varroa mite resistance was futile, "like breeding lambs resistant to wolves."[4] I referenced the findings of the Italian study, but it mostly fell on deaf ears. Beekeepers and scientists alike were applauding the development of miticides for use against this difficult pest.

Only Dr. Thomas Rinderer seemed determined to look for varroa mite resistance. He was a USDA bee researcher from Baton Rouge, Louisiana, who was studying the deleterious effects of Apistan, the miticide that was currently approved for use against varroa mites. He had researched its effects on honeybee longevity and queen fertility and found that fluvalinate (the active ingredient in Apistan) decreased drone longevity to below sexual maturity and that small doses greatly reduced queen fertility.

Because the varroa mite had come from Asia, Dr. Rinderer believed that mite-resistant bees might be found there. Our contingency from New Mexico urged him to investigate the matter, and eight years later he succeeded in importing Russian varroa-mite-resistant bees.[5] The Russian bees had adapted to the mite by learning to groom the mites off of one another. It is Dr. Rinderer's perseverance and confidence in mite resistance that has brought American beekeepers the strength of Russian genetics.

In 1995 I read that beekeepers in northern Mexico used creosote bush as a smoker fuel and had discovered that the smoke killed the varroa mites. They told Dr. Frank Eischen of the Weslaco, Texas bee laboratory about their discovery. He then conducted a study in which he tested the smoke of many different plant materials and found that some plant smokes killed both the mites and bees, but others, such as creosote bush, killed only the mites.

I had only two surviving hives that year and they were both top-bar hives. I decided to try using the medicinal smoke. I put a vegetable-oil-soaked paper on the floor of both of my heavily mite-infested hives and puffed creosote smoke into just one of them. I waited fifteen minutes. When

I looked into the hives, the hive that hadn't been smoked with creosote had ten to twelve mites on the paper and many mites on the bees. The hive I had treated with creosote had more than 300 mites on the paper and I couldn't find any mites on the bees. When I put creosote smoke into the second hive and waited fifteen minutes, I found more than 220 mites on the paper. We had found our crutch, a way to keep our bees alive until mite-resistant bees could be bred or imported.

I later found an IBRA (International Bee Research Association) publication, *Varroa! Fight the Mite,* that provided me with a detailed description of the life cycle of the mite. With this information, I came up with a plan that worked very well. I decided to cage my queens for ten days, so they couldn't lay any eggs. After ten days, I released the queens and waited seven more days. On the eighth day after the release of the queens, I began to smoke the hive with creosote bush and did so for seven days straight. Twenty-four days is the length of time that it takes drones to hatch, so this sequence of activities exposed all the mites in the hive to the medicinal smoke with the least amount of smoking.

I invited Dr. Frank Eischen to speak at the New Mexico Beekeepers Association meeting after he had published his plant materials paper.[6] He told us that he had been criticized severely for encouraging beekeepers to experiment with potentially dangerous and toxic herbal medicines. At the time, Apistan was the only approved substance to be used against varroa mites, yet the mites were already beginning to show resistance to it. Dr. Eischen explained the results of his study to our group and concluded by letting us know that he was unable to advocate legally for herbal treatments for mites because they currently were unapproved for use, but that there were no current regulations against various kinds of smoker fuels.

Eventually, I was able to buy Russian queens and no longer needed to use medicinal smoke. I have since established total resistance to varroa mites in my hives. Miticides were developed as the commercial medical response to varroa mites and, unfortunately, they still are used routinely by beekeepers to keep mites at bay. Miticidal strips are placed inside the hive and then are removed after a recommended period of time and disposed of as hazardous waste. The formula for miticides has changed over time in an ongoing response to the growing resistance of the mite to chemical applications.

Nosema

Nosema is a disease of the adult bee intestinal tract that gives bees extreme diarrhea. Bees are ordinarily very clean within the hive and will go outside to defecate. When a hive is infected by nosema, there will be streaks of yellow feces inside the hive as well as all over the entrance and front of the hive. Nosema can weaken a colony and eventually kill it. It is a fungus that is naturally occurring within many beehives, but in hives that are susceptible to an overgrowth of nosema, it can be deadly. It is commonly treated with antibiotics.

Recently, *Nosema cerranae* has gained notoriety as a potential contributing factor to Colony Collapse Disorder (CCD). However, many hives that have some amount of *Nosema cerranae* were not falling prey to CCD, while others were, so scientists were unable to point to it as a singular cause. While the results are still inconclusive in many areas of research, it is clear that bees are being compromised from many directions, and that a wide variety of symptoms and potential illnesses are the result. We feel strongly that a healthy hive can fend off *Nosema cerranae* easily, but that hives that are under attack from pesticide residues, the overuse of antibiotics and miticides, and the presence of other environmental pollutants are more susceptible to falling prey to this pathogen. [7]

Research also has shown that feeding bees honey that has been heated extensively while rendering wax also can cause a form of dysentery. It is best to dispose of heated honey rather than feeding it back to the bees.

Finding Good Genetics

As caretakers of more than one hundred hives, we have a large pool of resources when it comes to requeening. We keep careful notes on all of our hives, and when we observe that a queen is strong and disease-resistant, we can spread her genetics. Unfortunately, when a beekeeper is first starting out, it can be difficult to find mite- and disease-resistant queens. There is a good

FIGURE 8-2. A beautiful golden queen, raised without chemical treatments of any kind and selected for resistance to numerous diseases.

trend in beekeeping toward less treatment and more breeding for disease resistance and often conscientious queen breeders will advertise their use of IPM (Integrated Pest Management). Although this isn't a totally organic approach, it's definitely a step in the right direction and will yield potentially better results than the more conventional queen breeders.

Our best recommendation is to look for nucleus hives from beekeepers in the area who do not use any kind of inputs and who have been breeding their own bees for a number of years. The bees already will be adapted to the local ecosystem and typically will yield stronger and longer-lasting hives. When this option isn't available, catching swarms or removing hives from buildings and trees is also a way of obtaining local bees (see Chapter 3).

Packages of bees are usually ordered in the fall or late winter because suppliers like to ship the bees in the spring while the weather is still cool. It is best to order early because availability is becoming an increasing issue as honeybee populations continue to decrease. Be aware that breeders in the southern part of the United States may have genetic crossing going on with Africanized bees, so the bees you order from the South may be hotter than bees ordered from areas where Africanized bees cannot be found.

Symptoms of Insecticides

There is one affliction for which no amount of requeening or drugs can help.

I will never forget the day I came home to find all my bees dead and dying. I had been driving home admiring the billowing cumulus clouds building in the sky, but thoughts of the flowers that rain would conjure from the desert vanished when I opened my pickup door. The beeyard that should have been buzzing with hundreds of thousands of wings was terribly quiet. The silence drew my eyes to the hive entrances, and my dread turned to grief. There were piles of dead bees under every hive's entrance. On top of the piles were scattered bees stumbling around with their wings stuck at awkward angles. Inside the hives, I found dead and dying bees.

One of my neighbors had sprayed insecticide on his corn. When I tried to talk with him about the death of my hives, his response was hostile, "Do you think you can tell me not to protect my corn? Keep your #$!* bees out of my field!" It was a private property issue to him. Why should some beekeeper tell him not to spray his corn? To me it was a pollinator protection issue. Why should his decision to grow corn mean that all the pollinating insects in the neighborhood, including my bees, have to die?

In some cases, it's possible to convince others that they shouldn't use chemical sprays, but in the case of industrial agriculture, many farmers are being told how and what to farm by their chemical agents and are in debt to the system that they are being sold. Often bees simply have to be moved to safer areas.

If beekeepers want to protect their bees rather than move them, it is possible to get the state department of agriculture involved. Pesticide labels clearly state, "Do not use when bees are present," if they are toxic to bees. The label is a statement of law. If the poison is analyzed and it can be proven that someone nearby sprayed the insecticide at the time that the bees died, they may be liable for a fine. This means that the beekeeper has to pursue litigation with a neighbor. It is not easy to do, but bees increasingly need our protection.

Colony Collapse Disorder (CCD)

A different set of symptoms indicate a more insidious insecticidal death. When a hive dies due to Colony Collapse Disorder, beekeepers will find the hive full of honey, traces of brood, and perhaps a queen with a small handful of bees. The workers in the hive simply have flown away and never come home. Although the subject continues to be a controversial one in the United States, many countries already have banned a class of pesticides called neonicotinoids, finding them to be the cause of CCD. Minute traces of neonicotinoids have been found to have profoundly deleterious effects on the nervous and autoimmune systems of bees.[8]

Pollinators of every kind are at risk from this type of damage. There is a complex set of neurological activities that is required for a bee to navigate to and from the hive. A small amount of nerve damage, although it may not kill a bee, may effectively keep it from being able to pursue its livelihood.[9] It is the same with many other pollinators whose very existence is threatened by the continuing use of these kinds of chemicals. We can try to find places where there are little or no insecticides being used, but such places are increasingly hard to locate.

Our bees are like the canary in the coal mine, telling us when an area is too poisonous. As such, beekeepers are a small but increasingly noticeable rock in the shoe of modern agriculture. We can hope that highlighting the deaths of natural pollinators ultimately will force agriculture to find nonpoisonous ways to grow food. There are grassroots groups all over the globe advocating for pollinator-safe zones in which not only honeybees but all the many species of bees, wasps, butterflies, frogs, fish, birds, insects, and people can live safe from chemical pollution.

Unfortunately, as this book is being written, the debate is being reduced to dollars and cents, and the true value of honeybees is being lost in the equation. There are those of us who understand that if the honeybees and other pollinators disappear, there is little hope for any of us, but this is a weak sales pitch in the midst of the heightened market-driven environment in which we currently live.

As concerned beekeepers, we need to educate ourselves about the dangers that threaten our communities, and advocate against them. We once could trust in nature to provide for us, and as good stewards of these amazing creatures, we knew the returns were almost guaranteed. Now it can be very difficult indeed to care for honeybees, and beekeepers are becoming as threatened as their colonies. We strongly believe that advocating for organic agriculture is an important part of being a beekeeper in today's world.

Wax Moths

After I started keeping bees, the people in the town where I lived in began to refer to me as "the young bee man," and I started getting calls whenever anyone had a problem with honeybees. A local man had a honeybee colony established in the wall of his shed and wanted to get the bees out. He already had inquired with exterminators, but after finding out that it would cost several hundred dollars to have them poisoned, he decided to just leave them there. I agreed to take them out for him, and in the process I learned a lot about the hives of feral bees.

The hive in his shed had filled a whole wall. At the bottom of the wall was a healthy queen and all her attendants living on fresh white comb. Above them, in the middle of the wall, was slightly older comb stored with honey, and at the top of the wall, where they originally had begun building, was old black comb filled with wax moth larvae. The bees had abandoned these old combs and were coexisting with the wax moths, who were steadily munching away at the blackened remains that the hive had left behind.

Although we may not see them, there are wax moths, either latent or active, in all beehives. Pregnant moths have a scent-cloaking ability and slip into the hives past the guard bees and lay eggs. Tiny threadlike larvae then begin chewing their way through the combs, eating cocoon silk, honey, pollen, bee larvae, and beeswax. Wax moth larvae actually are unable to digest combs of clean, pure beeswax and instead thrive on old black combs filled with bee feces and layers of cocoons.

Sometimes we hear of hives succumbing to wax moth, but in many cases this is because all the combs in the hive simply had gotten too old and dirty to house bees, so the bees allowed the wax moths to take over. In a healthy hive, the bees in a colony are constantly weeding out wax moth larvae. If the hive gets sick or queenless and the population of bees diminishes, wriggling masses of wax moth larvae proliferate and destroy the hive. However, if a hive is healthy and strong, the bees can keep them at bay and even use them to remove old, unsafe combs. We consider wax moths to be symbiotic organisms in a beehive, like the wrecking ball that tears down an old condemned building to make space for new construction.

Whenever we harvest honeycomb that will be sold in plastic clamshells in its natural state, we store it for a time in the freezer to kill any wax moth larvae that may be present. If the honeycomb is left at room temperature, the larvae will hatch out and eat the honey. Storing honeycomb in the freezer has the added advantage of keeping it from crystallizing, so that it can be sold at any time of the year.

CHAPTER 9

Raising Queens

Beekeepers always are looking for the best bees. Even after more than thirty years of beekeeping, I am curious to try different strains of bees, so when I hear about queens that a breeder is proud of, I like to give them a try. Trying new strains often leads to some interesting discoveries and at the very least brings new genes to our genetic pool. However, the bees that inevitably perform the best in our geographic region are the ones that we breed locally. The resulting queens have genetics that are adapted to our unique set of environmental conditions, and they are able to mate with the feral bees in our area. Feral bees have demonstrated their survivability and often carry within them the genes for success.

FIGURE 9-1. A top bar that has been modified to hold queen cells. Two pieces of wood are customized to fit inside a top-bar hive, predrilled with holes to hold queen-cell cups, then hung with wire from the top bar.

Raising queens is a great skill for any beekeeper to learn because it strengthens and promotes local and sustainable beekeeping. Queen cells can be utilized to create new hives, to requeen hives that are struggling with poor queens, and to raise mated queens that can be stored in a queen bank for later use or for sale to other beekeepers.

Selection Criteria

Gentleness

Honeybees vary greatly in regard to their defensive behaviors. Each race of bees has its own unique characteristics, and there are also variations within each race. For instance, the Africanized bee is best known for its defensive behaviors and therefore has been labeled the *killer bee*. Yet there are regions of the world where this is the primary race of bees being worked with by beekeepers, and they are establishing gentle strains of African honeybees.[1] This is an effort that needs to be made with all races of bees, especially in a genetic melting pot in which many genetic profiles are being mixed with one another. Working with intensely defensive bees is not at all fun. Bees that sting readily, continue to sting long after the last hive visit, and follow and sting people or animals long distances from their hive can pose a threat to the beekeeper, the neighbors, and the public's acceptance of beekeeping. Therefore, gentleness is the first and perhaps the most important of the criteria for choosing hives from which to propagate.

It is important to realize that even if beekeepers choose their gentlest hive to raise queens with, it is possible for a queen to go out and mate with a drone from an aggressive hive, which then will produce aggressive offspring. The beekeeper has little control over this, but certainly can provide a second line of defense by eliminating any queens with aggressive offspring, thus preventing these hives from casting swarms into the surrounding area.

Disease and Parasite Resistance

Bees can thrive without any chemical support, but only if they show resistance to diseases and common parasites. Colonies that are afflicted with diseases or parasites are obviously not good choices from which to propagate. When considering which hives to raise queens from, it is important to consider which hives have consistently thrived and seem to be well adapted to the local environment. Russian bees, for instance, were imported to the United States because of their natural hygienic tendencies to groom one another of mites, making them resistant to infestations by the varroa mite. It is this kind of selection that provides the beekeeping community with strong bees. As attentive and careful beekeepers, we all can help to improve the overall health of our bees.

Honey Production

By keeping notes on honey production, we will see that certain colonies are responding well to the local climatic conditions and making the extra honey we desire. Often the hives that produce the most honey have the strongest queens and most desirable genetics. However, this is not always the case, and a strong honey-making hive may not overwinter well in the area or may have aggressive behavior.

Swarming Tendency

Sometimes in the springtime, during swarm season, a hive can be spaced and stretched, and swarming can be put off until the hive is almost full and can be divided. However, there are some hives that will swarm regardless of how much space they have in the hive. Their swarm tendency is high, which is an undesirable trait for the beekeeper, as it makes two small hives out of one and the resulting hives are much less likely to produce excess honey.

Swarm tendency varies from one strain of bee to another. Tropical bees, such as the African bee, are much more likely to swarm. In regions in which there are short winters or continuously warm weather, a hive doesn't need to store up a large reserve of honey due to an almost continuous supply of fresh nectar. Therefore, the bees can put all of their energy into reproduction. Conversely, bees that originate in regions in which the winter is long need to make sure they have a large amount of honey stored away to nourish them through the long cold months. They must resist the urge to swarm until the nectar flow is rich and sustained or they risk casting a swarm that has little chance of survival.

When we raise queens, we choose to raise them from beehives that begin raising swarm cells only after the hive is active and full of bees. This ensures that our resulting queens will be less likely to mother hives that are swarmy.

Color

Honeybees can vary in color quite a bit. Often the color of a queen correlates to her race. Italian bees are known to be light and golden in color, whereas Russian bees tend to be darker. Generally, the darker the bee, the more cold-hardy it is. Italian bees, although famous for their excellent honey-storage abilities, often have trouble overwintering in harsh climates, whereas Russians may yield less honey but will show greater survival rates in cold weather. When choosing a hive from which to propagate, the best choice will be the queen who has survived the most winters in the local environment, regardless of her color.

Creating Favorable Conditions for Queen Cells

When a colony of honeybees begins the process of swarming, it will start to raise a new queen. Beekeepers and honeybee research scientists have studied the factors that cause a colony to initiate queen-rearing so that they can reproduce these factors in order to raise an abundance of queens. The following factors are considered most influential in increasing the swarm urge in honeybees:

- The strength of the hive: Queen-raising requires copious quantities of royal jelly that in turn requires many young and well-fed nurse bees to be produced.
- The age distribution of workers: There is a higher percentage of young nurse bees relative to older workers.
- The flow of forage: Incoming nectar and pollen are plentiful.
- Congestion in the brood nest: There is a lack of space into which to expand.
- Low levels of chemical contamination in the hive and the surrounding area: Low levels of pesticides interfere with the bees' ability to raise strong, healthy queens.
- Queen pheromone production and distribution: There is low queen pheromone in the hive. New queens have a strong amount of

> pheromone, which suppresses swarming. As queens age, their pheromone production decreases and the hive is more likely to swarm.[2]

Beekeepers can induce and even enhance these conditions fairly consistently in order to raise queen cells.

There are also conditions that can be detrimental to raising queens: most notably the use of miticides has been shown to affect queen-raising adversely.[3] In areas of heavy pesticide use, contaminated pollen and nectar could degrade the hive's ability to raise healthy queens.[4]

There are several good books that sum up the queen-raising process, and beekeepers who are interested in becoming queen breeders should do some research and try various systems. The following is a description of our system and any adaptations that have been made for top-bar production.

Equipment

We like to use a minimum of equipment, but we do require a supply of queen-cell cups. There was a time when I made my own queen cups by taking rounded wet dowels and dipping them into molten beeswax. Although this is a very natural approach, it is a great deal of work, and the resulting cell is very delicate. The commercial queen cups that can be ordered from beekeeping catalogs are made of plastic and can be cleaned and reused, making them a good, long-lasting piece of equipment that is especially designed for our purposes.

These queen-cell cups need to be mounted on a top bar. Any top bar can be modified to become a queen-cell holder. On a typical queen-cell holder there are three rows of cells, one on the underside of the top bar and two rows mounted on pieces of thin wood that are hung from the top bar. Their length reflects the angled sides of the top-bar hive so that they can hang comfortably inside the hive without touching the sides. Holes are then drilled into the wood so that the queen-cell cups can be inserted snugly into each hole.

There are various grafting tools that are available commercially, but I prefer using a twig pulled from a tree. I have used twigs from apple, elm, and mesquite trees, but the best grafting tool I've found comes from coyote

FIGURE 9-2. Queen-cell cups that have been inserted snugly into predrilled holes.

willows. I prefer a slim, flexible, and pointed tip. If I pull the twig with a downward motion from where it originates on the main stem, it will pull off with a bit of skin and pulp from the main branch, which makes a perfect scoop for larvae. Because there is an unlimited supply of these handmade tools, they can be discarded once I'm finished using them.

Mating Nucs

The largest commitment that needs to be made to queen-raising is setting up the mating nucs. A *nuc* is a shorthand term that beekeepers use to refer to a nucleus hive; this is a small hive with combs of brood, honey, pollen, and a queen or, in the case of mating nucs, queens alternating with queen cells. The hives that are used for queen-raising often produce less honey due to the disruption of their brood cycles and their focus of energy on the production of royal jelly for new queens. The more queens we want to produce, the more hives we have to dedicate to that purpose.

When raising queens, it is helpful to have a small hive body or nuc, because once the queen begins laying, she will need to be found and caged, and the smaller the hive, the less time is spent searching for her. After a lot of experimentation, I found that simply inserting a partition into the middle of a top-bar hive provided me with two mating nucs of good size. Although this style of mating nuc is larger than the tiny mating nucs used in commercial queen-breeding, it is still considerably smaller than a regular beehive, yet large enough to grow to full size once I am finished raising queens in the spring. At that time, I simply transfer one of the two hives to another full-size box and pull the partition out, leaving the remaining hive room to expand or, alternately, the two hives can overwinter as mating nucs.

The small volume in the mating nuc means that the hive has to be managed carefully during nectar flows or it will fill up and become honey bound. Honey needs to be removed so that there is adequate room in the hive for the newly mated queen to begin laying eggs. The hive is large enough to have a strong field force, so it can store adequate amounts of honey to sustain itself during fluctuations in nectar flows. This means that the hive won't need to be fed during the queen-raising process. Once these nucs are going, they can maintain themselves and even become a source of several mixed-origin divides throughout the honey-making season.

Mating Nuc Placement

The mating nuc is the birthing hive for a series of virgin queens and will be the home they fly to and from for the few flights they make in their lives. If a batch of queen cells is grafted and put out on the same day, there may be several queens going out on their mating flights at the same time. Although they are escorted by field bees to the distant drone congregation areas and back, there can be confusion when relatively inexperienced flyers return to hives that are close together and have all the same color and orientation.

There is a phenomenon called *drift* in beeyards in which bees just learning to fly and orient themselves to their hive sometimes get confused on their first flights and come back to a different hive. They may drift slightly from the center of a beeyard to the corners or from the center of a line of beehives to the ends. In honey-production hives, this is a minor consideration, but in mating nucs drifting may mean a queen does not return to her hive and it becomes queenless. The queenless hive subsequently may drop into a depressingly low population and become a laying-worker hive.

Positioning the mating nucs at different angles, painting them different colors, and setting them at different corners of the beeyard helps the mating queens find their way back to the nuc they came from.

The Cell-Builder

There are two primary conditions that incite a vibrant and populous colony of honeybees to initiate queen-cell-building: one is the influx of lots of fresh nectar and pollen, and the other is a reduction or cessation of pheromones from the queen. We can cause the cessation of queen pheromones by removing the queen from a hive. But we must use careful timing for raising queens by waiting until a fairly dependable bloom is under way and the bees are filling the hive with honey. When the hive begins to fill with honey and brood, the brood nest will start to feel crowded and the hive will begin to feel the need to swarm.

The first sign of the swarming urge is capped drone brood. Capped drone brood is a signal from the bees that resources are plentiful and that they are getting in the mood to raise queens. There is little use trying to get bees to raise twenty or so good, heavy queens if they are not even trying to raise drones. In commercial breeding environments, bees often are fed with sugar syrup and pollen substitutes to create this condition, but we wait for all the natural indicators to be present. Even if we could incite the bees to swarm artificially, there is no guarantee that there would be drones for them to mate with unless hives in the surrounding areas also were being artificially compelled to swarm. Bees thrive through outcrossing, so when we raise queens, we need to be sure that there are ample drones in the natural environment to bring diversity into our genetic pool.

In natural top-bar beekeeping, drone comb construction is not inhibited and when the bees have decided to raise drones, they will construct drone comb and direct the queen to fill it with eggs, providing us with our first natural indicator that the time is ripe for propagation.

Setting Up the Graft

When preparing to graft, it is necessary to calculate how many queens are being raised and how many mating nucs will be needed to house queen cells. Especially in the beginning, when one is beginning to experiment with this process, some of the grafts will fail, so it is good to shoot for more rather than less. Often we will place out more than one queen cell per mating nuc, so that the hive has two opportunities to hatch a queen. If I have ten mating nucs available, then I might graft twenty-five larvae, with the hope of getting twenty viable queen cells.

In order to begin the process of queen-raising, I need to open my mother colony or the hive that will be the mother of all my future queens. As I go through this hive, I select one comb that will donate its larvae and mark it, leaving it in the mother hive until I am ready to use it. This comb is chosen because it has a large number of eggs and very young larvae in worker-sized cells.

Worker larvae that are less than a day old are the best candidates for raising queens because they have not yet been differentiated into workers and can still evolve into queens under the right conditions. Larvae more than a day old may end up as intercaste queens, having been turned into queens after they had already begun the differentiation process into workers. Honeybee eggs, although young enough for queen-raising, are very delicate and cannot be lifted and transferred to queen-cell cups and kept consistently viable. Drone-sized cells are not a source of larvae for queen-raising because they contain only male larvae that cannot be raised into queens.

My next job is to prepare the cell-building hive. My goal is to take a hive that was teeming with bees on twelve to fourteen combs and crowd all of those bees onto eight or nine combs, while removing all of their eggs, young larvae, and their queen. In order to do this, I open six or seven hives in the beeyard and leave them open at the back of the brood nest so that they can receive donations of broodcomb from the cell-building hive. Then I remove

FIGURE 9-3. A young larva being lifted out of its cell to be placed into a queen cup. The tool being used is a willow twig.

any broodcombs with eggs or young larvae from the cell-builder, brush off the bees, and distribute the combs to all of the open hives in the yard.

The uncapped brood is distributed to a large number of hives because no single hive can receive a large supply of brood without being overtaxed. Most hives operate at their maximum level of expansion already, and if we were to distribute multiple combs of uncapped brood to one hive, some of it most likely would be uncared for and would perish. Lastly, I remove the queen and either cage her for future use or place her in a queenless hive. She even could be placed in a queen bank until the cell-building is complete and then returned to the cell-building hive.

By removing most of the hive's young brood and their queen, I am taking all of the energy of the hive and refocusing it on my queen-raising process. The bees no longer have eggs or young larvae to feed, and the capped brood remaining in the hive soon will yield nurse bees who can contribute to the queen-raising process. The bees had been engaged actively in producing large quantities of brood-food for the young larvae, and now instead they will be engaged in producing royal jelly for the twenty-five queens I want to raise.

Once the cell-building colony has been effectively *crowded*, the larvae-supply comb from the mother hive is moved in (without any bees from the mother hive) and so is the top bar with the empty queen-cell cups hanging from it. The graft will not be conducted until the next day, when the hive has fully understood that it is queenless, has cleaned and prepped the queen cells it has been given, and is ready to begin the process of raising a quantity of new queens. The presence of the larvae in the supply comb keeps the females from beginning to develop their ovaries as laying workers and it also keeps the morale of the hive from dropping into a mood of desperation. During the night, the bees will clean, disinfect, and coat the plastic queen-cell cups with propolis and wax.

Before being closed up for the night, the cell-building hive should be assembled carefully in preparation for the graft. The combs in the hive are arranged with the following pattern, starting with the comb nearest the entrance:

1. A comb of pollen and honey
2. An empty top bar or a top bar with a partially built comb. The purpose of this empty bar is to absorb the comb-building desire of the colony and to prevent them from building too much comb on the bar holding the queen-cell cups.
3. A comb with honey, pollen, and a small amount of brood
4. Another comb of honey, pollen, and a small amount of brood. It is important to have some combs with pollen near the cell bars because the nurse bees will be digesting a lot of pollen to make royal jelly for the queens.
5. The top bar with the queen-cell cups
6. The larvae-supply comb from the mother colony, which should be marked for easy retrieval
7. Another comb of honey, pollen, and a small amount of brood
8. A comb of honey
9. An empty top bar or top bar with a partially built comb. This bar serves the same purpose as the second top bar.

The cell-building colony now can be closed up and left quiet for twelve to twenty-four hours, when it will have developed its maximum brood-food surplus. The brood-food surplus happens because of the lack of young brood now present in the hive. The brood-food surplus will not reach its peak until twelve hours after the young brood is removed and then, after twenty-four, the hive will begin to absorb brood-food and reduce its production. If the beekeeper gently places larvae in the warm, propolized queen-cell cups during this window, when brood-food surplus is at its peak, there is a good chance the bees will raise them into great queens.

The Graft

Grafting is the term used by beekeepers to describe the act of taking a young larva, removing it from its beeswax cell, and placing it in a queen cup. In laboratory environments, this sort of work is done with a microscope. When doing this in the field, it can be helpful to have reading glasses or a handheld magnifying glass, but it is also quite possible to do this work with the naked eye.

Grafting in the morning, when the humidity is high, helps to keep the brood fresh. In arid climates, it also can be helpful to have a spray bottle filled with clean water and a wet towel to keep the air around the comb humidified because the larvae can dry out fairly quickly without the protective presence of nurse bees. The towel is used to rest the comb on while scooping larvae. Although we want the broodcomb and the queen-cell bars to be humid, we also want to avoid getting water on the brood and in the queen-cell cups, so the water bottle should be sprayed at an angle to keep water droplets from getting into the bottoms of the cells.

To begin the grafting process, the cell-builder hive is opened with very minimal disturbance. All that is needed is to crack open the hive to retrieve the fifth and sixth bars (the top bars with the queen cells and the larvae-supply comb). Hopefully, a thick coating of bees is festooned on them both, indicating that there is a strong population in the hive. The bees need to be brushed off both the cell bar and the larvae-supply comb, and then these two bars should be taken to a comfortable spot away from the hives because this work will need to be done without a veil.

Both the cell bar and the larvae-supply comb are placed on the wet towel and given a quick mist before the larvae-lifting begins. Then the grafting tool is used to lift one larva at a time into the queen-cell cups. The best approach

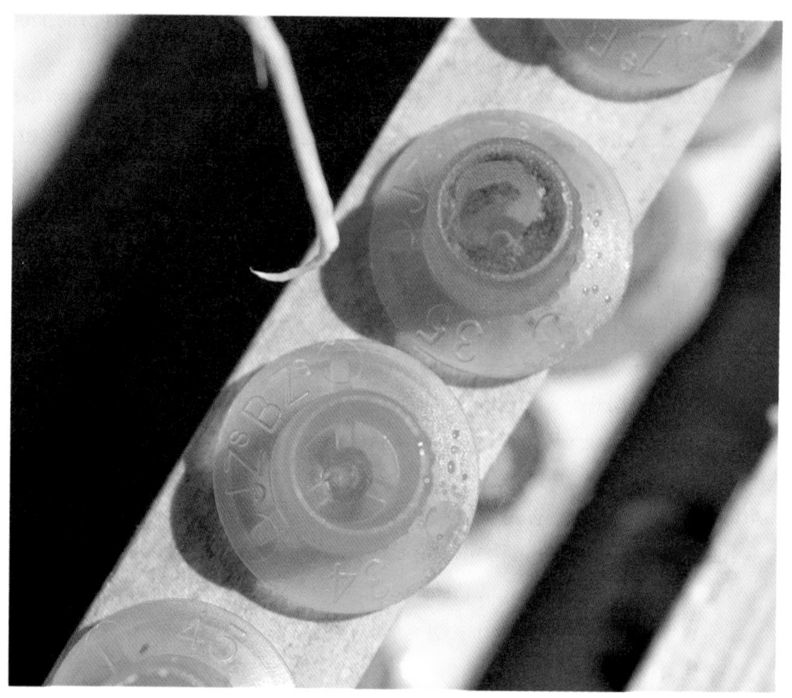

FIGURE 9-4. The young larva is transported quickly to the queen-cell cup and gently laid in the middle. A light mist of water is used to keep the surrounding air humidified so the larva will not dry out during this process.

is to slide the tip of the grafting tool along the bottom of the cell behind the C-shape of the larva, so that either end of the larva is hanging slightly off of the tool as it is lifted. The more royal jelly that can be scooped up with the larva, the better. Once it has been placed in the queen-cell cup, the royal jelly will keep the larva moist and well fed. When removing the larva from the beeswax cell, it is important to keep it from touching the sides of the cell because it will cling easily to the side and slide from the grafting tool. The larva then is deposited in the bottom of the queen-cell cup by tipping the grafting tool gently to the side. Mist the queen cells and larvae-supply comb every couple of minutes until every queen-cell cup has a larva grafted into it.

When choosing larvae to graft, it is important to pick the smallest larvae available, and no larva greater than four times the size of an egg is suitable. Larvae that are too big have already developed worker characteristics and will make poor, small queens. It takes practice to develop an eye for the right size of larvae and even more practice to graft effectively. In the beginning, many of the grafts will fail, but some will survive and, given time and practice, this ratio will shift until most of your grafts will be viable.

As soon as the grafting is complete, the queen-cell cups need to be placed back in the cell-builder hive so the bees can begin to care for them. The larvae-supply comb should be checked for queen cells before it is placed back into the mother hive because often the bees in the cell-builder will have begun the queen-raising process already with some of its younger worker larvae. These queen cells should be broken down before returning the comb to its original hive.

The graft is then left overnight in the cell-building nuc and checked the next day. If the graft was successful and there are viable queen larvae, the queen-cell cups will have bell-shaped beeswax cowlings built onto them and they will be full of royal jelly. The grafts that were not successful will not show any signs of wax building and the cells will be empty. This is an opportunity to refill the cells with fresh larvae. If less than 50 percent of the queen-cell cups do not contain viable larvae, it may be that the cell-builder is not strong enough to raise a large number of queens and a new cell-builder needs to be chosen. A successful graft is one in which 80 percent of the cells are viable. Once the graft has been evaluated, the cell-builder hive is closed up and left alone for a week.

FIGURE 9-5. A beautiful batch of queen cells, capped and ready to be put into their mating nucs.

Exactly a week later, the graft is checked again to see that all of the queen cells are capped. Any queen cells that are particularly small or uncapped are culled at this time because they are not likely to produce good queens. The queen cells must be placed into their mating nucs before ten days have passed or one of the queens will hatch and kill all of the others while they are still in the cell-building hive. Instead of having twenty or so new queens, there will be only one, and the work will have been wasted.

When the graft is checked to see if the cells have been capped, a count of the cells is taken and the appropriate number of mating nucs are prepared. It is possible to put in one or more cells per nuc. If more than one cell is placed in a mating nuc, the queen that hatches first will end up killing the other queen(s). If only one cell is put in each mating nuc, and that cell is not viable, the mating nuc will end up queenless. If there is any question about the viability of the cells, putting two or more cells per nucleus hive can help ensure that each mating nuc ends up with a queen. Often when beehives are preparing to swarm, they will raise five to eight queens, giving themselves a greater natural opportunity to receive a good queen.

The first step in preparing the mating nucs is to remove their queens. The hives that are chosen should have plenty of bees and fresh brood. The

FIGURE 9-6. The queen cell is removed by grasping the sides of the cup and gently pulling down. The cup then is pushed into the top of a broodcomb, taking care not to put any pressure on the outside of the delicate queen cell.

queens are caged and put into a queen bank until they can be given a permanent home. This process is described later in this chapter.

The second step is to leave the now queenless mating nucs open so that they can receive a queen cell. Always choose the same spot in each hive so that it is easy to check later. I place queen cells between the third and fourth top bars, making sure that there is brood on both of these combs. Once all of the mating nucs are open between the third and fourth bar, I retrieve the cells from the cell-building hive and begin to place them in the mating nucs.

When removing the cells from the cell-building bars, it is very important to be extremely gentle. The queen inside the cell is still pupating and is very delicate. Shaking or jarring her can make her unable to fly or otherwise damage her and render her useless as a queen. The cells are pried gently from the bar and placed one by one into each of the mating nucs. The plastic base is pressed into the wax near the top of the third bar, with the queen cell hanging down, in the same position as it was in the cell-building hive. The third and fourth top bars then are pushed gently back together and the mating nuc is closed up.

Managing Mating Nucs

The most limiting factor in the queen-raising process is the strength and number of the mating nucs. The mating nucs are a limited resource because their queens are being taken from them constantly and therefore the number of bees inside them is diminishing continuously due to a lack of new brood. We remove their original queens and give them cells that are about to hatch. When they hatch, the virgin queen is fed, and it takes five to twenty days (depending on the weather and hive resources) before she leaves for her mating flights. The mating period can last several days, and then once she is mated she spends a few days adjusting to pregnancy before she starts laying eggs. The worker population of the nuc diminishes until the new queen's eggs begin to hatch, twenty-one days after they were laid. If the mating nuc is going to be reused for another batch of queens, then this queen will be removed and yet another queen cell will be placed inside, and so on. Eventually, the mating nuc runs out of worker bees to support the process.

Mating nucs need to be monitored to make sure they are maintaining the strength they need in order to participate successfully in the queen-raising process. If a mating nuc is in need of assistance, combs of capped brood from support hives may be donated to it, providing it with additional nurse bees. Alternately, the newly laying queen can be left in the hive long enough to provide it with a replenishment of brood before she is removed and a new cell is placed inside. Often I let the newly mated queens lay eggs for ten to fourteen days so that the mating nuc can build itself up before the next round of cells is placed out. I prefer to have the mating nuc support itself rather than pulling brood from honey-making hives. This period of time also provides me with an opportunity to evaluate the queen's egg-laying capacity adequately.

Caging Queens

A queen cage is a little screened container used to contain queens for periods of transport or temporary storage. If the queen is being held in the company of workers, she is usually inside the cage alone. If she is being transported without a cluster of workers around the cage, some workers are put into the cage to care for her. Because she cannot feed herself, the queen will die without workers present. The mesh of a queen cage is always large enough to allow worker bees to feed the queen through the holes. There are several types of cages available from beekeeping supply companies.

Queen cages are handy to have around, regardless of whether one is raising queens or not. When we are doing hive removals or catching swarms, finding and caging the queen can ensure the success of the project. It also can be helpful at times to cage a queen in the beeyard if she is being moved from one hive to another.

Most queen cages have two entryways, one to put the queen and attendants in and one that can be filled with queen-cage candy. Queen-cage candy can be made at home with two parts honey and five parts powdered sugar, by weight. If the queen is being caged in very hot weather, the amount of powdered sugar can be increased to six parts for a firmer candy. The powdered sugar should be cane or beet sugar because corn sugar contains too much starch. High-fructose corn syrup is not recommended because it contains chemicals that can be harmful to the bees. The mixture is kneaded until it is the consistency of stiff dough. It is then pressed into the tube or hole before the cage is ready to receive the queen.

A useful skill for beekeepers to develop is the ability to pick up bees with our fingers without getting stung. Caging queens requires this skill, and we become all the more useful to family and friends if we can toss bees out of cars and homes easily. To pick up bees, gently grab both sets of wings between the thumb and forefinger. If both wings are being held, the bee's thorax is between the thumb and forefinger and its legs, abdomen, and stinger are pointed away from the fingers so that it cannot sting. Learning to pick up bees by their wings takes practice, and we often spend time in classes doing this simple task so that students develop a comfort level with handling bees. Drones are easy to practice with because they have no stingers.

To cage a queen, gently grasp her wings and remove her from the hive. Then gently position her head, legs, and thorax into the entrance of the cage. It is very important to avoid hurting her in any way, so take as much time as is needed to ensure that she isn't scraped against the sides of the cage. Put a gentle pressure on her if she tries to back out and release her if she goes forward, giving her no choice but to walk into the cage. If some workers need to be put inside the cage with her, use a finger to close the opening until enough bees are inside the cage and then close the plug.

When the workers are in the cage with the queen, the cage is easy to carry around as an independent unit. The workers will eat the candy and feed the queen, and a small amount of water can be provided in little drops on the surface of the cage. It is important to keep the cage out of direct sunlight. Often I will carry queens around in my shirt pocket, although at certain times of the summer, when it is very hot, the queen can die of heat and a lack of air circulation. A shady, moist environment is the best for storing queens until they can be rehived.

If the queen is being transferred to a new hive, the presence of the workers in the cage may reduce the likelihood of her being accepted, so it is better not to include any workers in her cage. She cannot live for long without workers to feed her, so the transfer must be rapid. By the time the workers in the new hive have eaten all of the sugar candy, they will have adapted to her

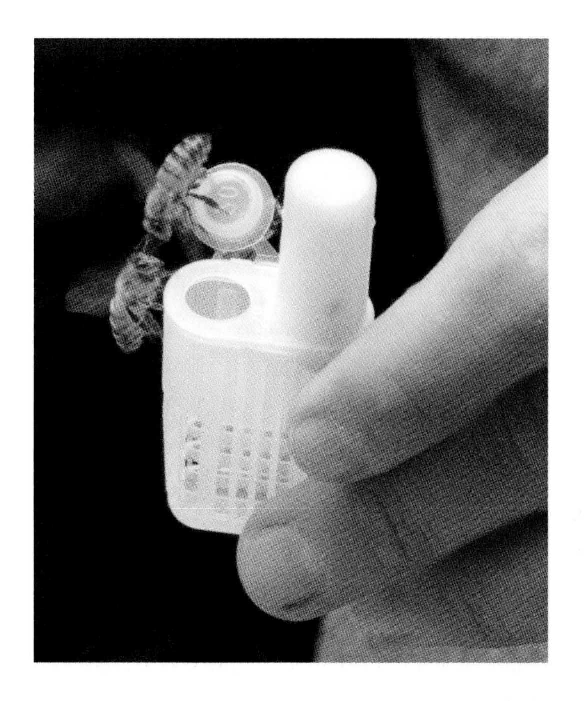

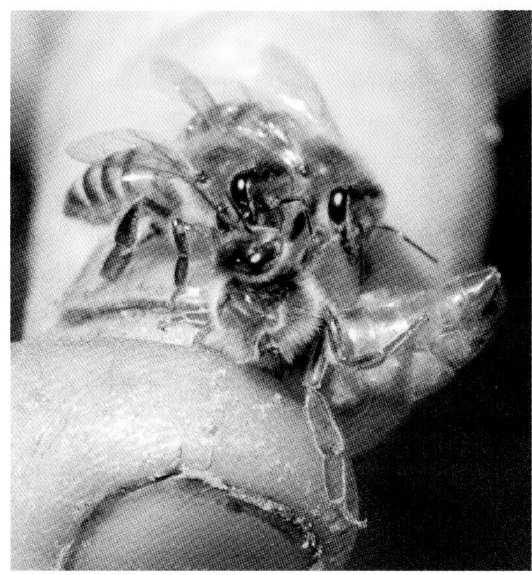

FIGURE 9-7. A queen-cell cage (top) and a queen being held by her wings. Learning how to hold and cage a queen takes practice. Stingless drones are good for this purpose.

scent and will accept her into their community. If there are workers with her, she seems more like an invading force and is more likely to be killed by the new hive when she is released.

Because workers have a short lifespan, a worker may die inside the queen cage before the queen is released. The dead body can trap or tangle the other bees or plug the doorway so they cannot get out when it is time to leave. In cases like this, it is possible to release the queen in a contained environment and then either put her back in the cage, minus the dead worker(s), or put her into her new hive. A bee veil provides an excellent environment for doing this sort of work. The dead bees can be taken out of the cage, and if the queen gets out as well, she will be easy to get back into the cage.

Queen Banks

When caging and storing multiple queens for a period of time, it is helpful to have a queen bank. A queen bank can be a queenless hive full of workers, honey, and pollen in which the queens can be stored indefinitely until we are ready to place them, or it can be a mobile box that contains a secure chunk of honeycomb and pollen and a number of workers to feed the queens. Workers naturally will gather around the queen cages, so simply leaving a queen-cage box open around some hives will draw a number of workers into the box, at which time it can be closed, trapping them inside. There should be approximately ten workers for each queen. If there aren't enough in the box, gently brush some in from a nearby hive. Having a number of workers loose in the box avoids the need to cage them with the queens.

A mobile queen bank can be maintained for weeks or even months but requires new bees and food every three to four days. There are boxes manufactured just for this purpose. They have proper ventilation and are slotted to keep the queen cages from moving around during transport. Queen banks should be kept out of the sun and often are stored indoors where there is little temperature fluctuation. Water is brushed onto the outside of the box once a day or more, where the workers can gather it through the ventilation holes. Queen banks shouldn't be left in hot or cold vehicles but are better carried around with the beekeeper if he or she has to stop for a meal or a meeting. Many interesting conversations have come from such situations because the bees make a loud buzzing when they are being moved around. If the queens need to stay in the bank longer than four days, the box should be opened in a beeyard to let the trapped bees out (they will be accepted by the hives) and new bees should be brushed in. New honey also can be added at this time if needed.

When creating a queen bank from a hive, take two or three combs of honey and pollen and two or three combs of capped brood and transfer them into an empty box. Make sure there is no queen present and no eggs or young larvae with which the bees could raise a queen. Place the caged queens between the combs so the bees have easy access to the queens for feeding. In this environment, the queen-cage candy is covered with a cap, so that the bees cannot release the queens. In a few days, check the bank for queen cells and remove or relocate any that are found. This hive can be maintained for a long time, even several months if needed, by adding two combs of capped brood every two weeks and removing any queen cells as they occur. It is better not to cage queens for very long, though, because they can die or suffer from their containment.

CHAPTER 10
Planting for Bees

There are many ways in which we can support the world of honeybees, but in our opinion one of the best ways is to develop diverse habitats for them to live in. Any homeowner or gardener can add some plants to the landscape, providing an important food resource to these wonderful insects. Often plants that benefit honeybees also attract hummingbirds, butterflies, and other important pollinators. Flowers bring not only life but also beauty in their diverse colors and forms.

Pollinators are at risk mainly due to diminishing habitats and are falling prey to diseases that may be based on symptoms of malnutrition. Wild landscapes provide a diversity of nectar and pollen throughout the season, whereas monocropped farmlands are a virtual desert of forage for most pollinators except when they are in bloom, at which time they provide only a limited form of nourishment.

We serve pollinators the most by providing them with a wide variety of plants to feed from over an extended period of time. The following lists of species are provided purely as suggestions and were found by doing extensive research for our particular ecosystem. When researching plants for pollinators, it is necessary to look at planting requirements and zone hardiness. We have made our choices based on a semidesert, high-altitude environment, and these plants may or may not be appropriate in other ecosystems. When purchasing plants or seeds, it is also important to consider how difficult the plant is to grow and how much labor is required to get it established.

FIGURE 10-1. The honey cycle. Illustration by Christopher Clow

Within this list of plants, we've included trees, shrubs, perennials, annuals, and cover crops. All of these are useful for pollinators, but trees are often the best investment, especially when space is limited. A linden tree planted in an ideal location and watered regularly can yield a huge amount of nectar once it is mature. In order to provide some sort of nourishment throughout the season, however, it is helpful to have some plants in each of these categories.

Trees

FRUIT TREES

Fruit trees are a wonderful source of nectar and pollen for honeybees and other pollinators in early spring. Honeybees will help to increase yields, so there is mutual benefit to the bees and the homeowner. It is important to choose varieties that actually require pollination. Many ornamental varieties of fruit tree are self-pollinating and have little attraction for honeybees. Often older heirloom or native varieties are most beneficial for bees. Apples, cherries, and plums are particularly attractive to bees and make excellent early spring forage.

WILLOW

Willow trees provide bees with early pollen, jump-starting the hive with a valuable source of fresh protein in the spring. There is a large variety of willow trees and shrubs available, many of which are ornamental and can be harvested for use in basketry or planted as hedges or windbreaks.

HONEY LOCUST

There is a wide variety of locust trees available for planting. Typically, locust trees bloom fairly early in the season, providing an important source of early nectar and pollen. The honey from these trees is delicious!

CATALPA

Catalpa is also a fairly early-season bloomer and another delicious source of honey.

LINDEN

Linden trees bloom in midsummer, making their nectar flows more reliable than spring-blooming trees, which can lose their blooms to late frost. A horticulturalist we know referred to this tree as a "bee restaurant." A mature tree in full bloom is a truly lush meal for a variety of pollinators. It also provides a rich, buttery honey that is totally delicious.

BEE BEE TREE

Also known as the Korean Evodia tree, this tree is famous for its attractiveness to bees, hence its common name of bee bee tree. Also a midsummer bloomer, it can be planted as an individual tree or used in groupings as a hedge or windscreen.

TARTARIAN MAPLE

This particular maple makes small white flowers that the bees adore. The flowers are followed by the winged seeds that helicopter to the ground in fall.

OTHER TREES

There is a wide variety of trees that provide forage for bees; only our favorites are mentioned above. Tulip poplars and various ash trees are also good, but the best way to find the best trees for a particular area is to visit nurseries and extension agencies, with a particular question in mind: Do they attract bees and pollinators when they bloom?

Shrubs

PEASHRUB

This shrub has leguminous flowers that are attractive to bees. Peashrubs commonly are used as windbreak plantings and are beneficial because they fix nitrogen in the soil.

COTONEASTER
This is a shrub that offers clusters of flowers that turn into berries. They also are good for windbreaks and wildlife plantings.

CHOKECHERRY
Chokecherry can be used as a shrub or tree and has long panicles of white flowers followed by berries. It also is good for windbreaks and wildlife plantings.

VIBURNUM
There are a number of different viburnum species that are beneficial to bees.

LEAD PLANT
This is a drought-tolerant shrub that is valuable as a forage species.

SPIREA
This ornamental plant is very attractive to bees and is available in a number of blue and purple varieties.

LAVENDER
A classic honey-producing flower, lavender is drought-tolerant and a prolific bloomer.

NEW MEXICO PRIVET
This is a good early source of nectar for riparian areas.

Perennials

COLUMBINE
Attractive to honeybees and hummingbird moths, columbine is a good early source of nectar in the mountains.

COMFREY
Well loved by honeybees and bumblebees alike, comfrey is a repeat bloomer that is also an excellent ground cover and green manure.

MINT
There are a number of varieties of mint that are attractive to bees: basils, catmint, thyme, oregano, peppermint, licorice mint, lemon balm, and others.

ECHINOPS (GLOBE THISTLE)
This is an ornamental thistle plant that has a beautiful, steely blue globe flower that honeybees love.

ALLIUMS
The Alliaceae, or onion, family is well loved by bees. These bulbs are rodent- and deer-resistant, and some varieties will spread willingly.

ANGELICA
Also used as a medicinal herb, this large flowering plant is very attractive to a number of different pollinators.

MOTHERWORT
Another medicinal herb that is a very prolific bloomer, motherwort is easy to grow, is somewhat invasive, and is a great plant for neglected areas.

ST. JOHN'S WORT
Another medicinal herb, it is also a drought-tolerant species and is easy to grow. Pretty yellow flowers are used to make oils and tinctures and are attractive to bees.

MALLOW

There are a number of species of mallow that the bees love. We have a native orange globe mallow, and there are various other perennial species, such as marsh mallow, Mauritanian mallow, and hollyhock. These offer pollen as well as nectar.

ROSES

There are some wonderful roses that bees love, but they are similar to fruit trees in that the variety makes a difference in the nectar and pollen content. Many of the old-fashioned, single-petaled varieties are the most useful to bees.

HYSSOP

The flowers often are used in tea and are a good source of nectar in the heat of summer. Hyssop is somewhat drought-tolerant, similar to lavender.

SALVIA

Salvia is similar to lavender and hyssop, with blue spires of flowers that bees enjoy.

MONARDA

Also commonly known as bee balm, this plant is most attractive to butterflies and bumblebees.

Cover Crops

SWEET CLOVER

There is hardly a plant that honeybees love more. Yellow sweet clover is a self-seeding annual, and white sweet clover is a biennial.

ALFALFA

Alfalfa is a great source of summer nectar that yields a dark amber honey.

ALSIKE CLOVER

This is another nitrogen-fixing plant that honeybees love.

BUCKWHEAT

Buckwheat is great for weed suppression and yields flowers in as little as twenty-five days. When this plant is in bloom, it simply hums with activity from honeybees, ants, and wasps.

YELLOW MUSTARD

This is a plant that honeybees love, and it grows wild in many areas, but seed is also available for planting as a cover crop.

PHACELIA

This plant is used a great deal in Europe and is famous for yielding a very clear, white honey. There is a variety that grows wild in southern New Mexico, but it also can be obtained as cover-crop seed and will bloom continuously until frost.

HAIRY VETCH

This is another nitrogen-fixing plant with pealike, leguminous flowers.

Annuals

BORAGE

Honeybees love this plant. It provides a continuous supply of little periwinkle-blue flowers throughout the summer. The flowers and young leaves can be eaten in salads. In moist conditions this plant will volunteer happily each year.

SUNFLOWERS

Sunflowers are a great source of pollen and nectar for honeybees.

COSMOS

Cosmos is another pretty flower to brighten up the garden that bees also enjoy.

CLEOME

This plant is referred to commonly as spider plant.

TULSI BASIL

In the mint family, this plant offers continuous bloom throughout the summer. Honeybees love it, and it can be harvested and dried at the end of the season to be used as a winter tea.

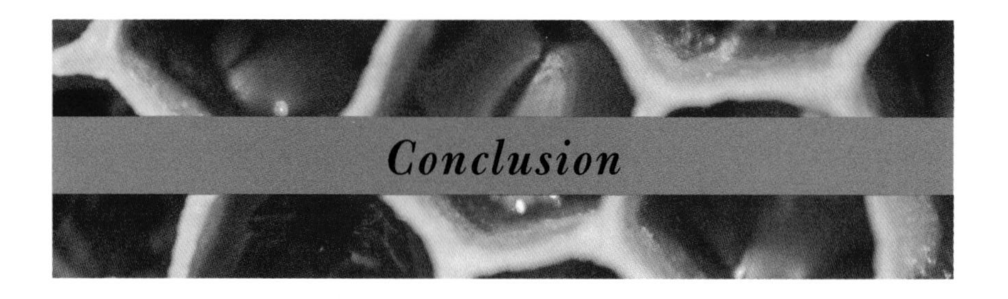

Conclusion

As beekeepers, we are intimately tied to our environment and the symptoms of its degradation. In the past century, beekeepers could rely on the natural abundance of Mother Nature, in which we could always find some reliable measure of support. Those times have changed, and our roles have shifted. What we need now is a model of cooperation and symbiotic co-creation. Without pollinators, our current population cannot survive, and similarly, without our support and protection, pollinators cannot survive.

It used to be said that beekeepers couldn't plant enough flowers to make a difference to the bees, but in these days of shrinking wildlife habitats, tremendous losses in biodiversity, and invisible onslaughts from pollution and man-made insecticides, we have no choice but to try. In the world we find ourselves in today, no garden and no prayer of support for nature is too small. No measure of effort is too little from those of us who have a mind and spirit to care about our pollinators enough to plant some food to feed them. We no longer have the luxury of being passive spectators at a feast at which we are the honored guest. It's time to invite nature to our table and to honor her with a bountiful repast. With a strong collective effort, we can all participate in giving birth to hope for our unique community of honeybees and people.

In our lifetime, we've watched astronauts walk on the moon. That remarkable feat was a demonstration of the heights to which our technology and determination can take us. Yet the base of our immense technological tower has its foundations in the biosphere of our one and only planet. We risk forgetting that we cannot reach to such great heights without the incredible web of life that supports us. Honeybees are one of the great contributors to our success, and yet they are rarely noticed or given the respect and gratitude that they deserve.

The year this book was written (2011) saw massive declines in the populations of bumblebees, monarch butterflies, and numerous other pollinators. Honeybees are just one among many of the insects that are under threat of potential extinction. We cannot continue to defer the issue. The time is now to shift our perspectives and make a commitment to supporting life as we know it with every action we can make.

In a way, the term *beekeeping* is a misnomer. The door to a beehive has to be open in order for the hive to live through a day. We don't keep our bees; we give them a place to live and do our best to protect them, and they do the same for us. Let's do this together, and with our hearts and minds as one, we can see our pollinator populations flourish once again.

Notes

INTRODUCTION

1. Steve Taber, "Determining Resistance to Brood Disease," *American Bee Journal* 122(1982): 422–425.
2. M. Spivak, "Honeybee Hygienic Behavior and Defense Against Varroa Jacobsoni," *Apidologie* 27(1996): 245–260.

CHAPTER 1

1. B. Smedal, M. Brynem, C. D. Kreibich, and G. V. Amdam, "Brood Pheromone Suppresses Physiology of Extreme Longevity in Honeybees (*Apis Mellifera*)," *Journal of Experimental Biology* 212 (23) (2009): 3795–3801.
2. Giancarlo A. Piccirillo and David De Jong, "The Influence of Brood Comb Cell Size on the Reproductive Behavior of the Ectoparasitic Mite Varroa Destructor in Africanized Honey Bee Colonies," *Genetics and Molecular Research* 2 (1) (2003): 36–42.
3. M. Kubik, J. Nowacki, A. Pidek, Z. Warakomska, L. Michalczuk, and W. Goszczynski, "Pesticide Residues in Bee Products Collected from Cherry Trees Protected During Blooming Period with Contact and Systemic Fungicides," *Apidologie* 30(1999): 521–532.
4. Judy Y. Wu, Carol M. Anelli, and Walter S. Sheppard, "Sub-Lethal Effects of Pesticide Residues in Brood Comb on Worker Honey Bee (*Apis Mellifera*) Development and Longevity," PLoS ONE 6 (2) (23 February 2011), e14720, www.plosbiology.org.
5. E. R. Jaycox and S. G. Parise, "Homesite Selection by Italian Honeybee Swarm, *Apis Mellifera ligustica* (*Hymenoptera: Apidae*)," *Journal of Kansas Entomological Society* 54(1981): 697–703.

CHAPTER 2

1. E. Southwick and R. Moritz, "Social Control of Air Ventilation in Colonies of Honey Bees, *Apis Mellifera*," *Journal of Insect Physiology* 33 (9) (1987): 623–626.

CHAPTER 3

1. P. K. Visscher, R. S. Vetter, and S. Camazine, "Removing Bee Stings," *Lancet.* 348 (9023) (1996): 301–302.

CHAPTER 4

1. Giancarlo A. Piccirillo and David De Jong, "Old Honey Bee Brood Combs Are More Infested by the Mite Varroa Destructor Than Are New Brood Combs," *Apidologie* 35(2004): 359–364.
2. "The Effect of Old Dark Comb in the Beehive," a presentation to New Mexico Beekeepers Association, 1984, by Dr. Elbert Jaycox, honeybee researcher from University of Illinois.
3. Blaise W. LeBlanc, Gillian Eggleston, Diana Sammataro, Charles Cornett, Renee Dufault, Thomas Deeby and Eldwin St. Cyr, "Formation of Hydroxymethylfurfural in Domestic High-Fructose Corn Syrup and Its Toxicity to the Honey Bee (*Apis Mellifera*)," *Journal of Agricultural and Food Chemistry* 57 (16) (2009): 7369–7376.

4. "The Microflora of the Honey Bee," a presentation to the New Mexico Beekeepers Association, 2009, by Dr. Diana Sammataro, honeybee researcher.
5. Martha Gilliam, Carl Hayden Bee Research Center, "Microbiology of Pollen and Bee Bread: The Genus *Bacillus*," *Apidologie* 10 (3) (1979): 269–274; Gloria DeGrande-Hoffman, Russell Vreeland, Diana Sammataro, and Ruben Alarcon, "The Importance of Microbes in Nutrition and Health of Honey Bee Colonies," *American Bee Journal* (2009): 755–757.
6. Luísa G. Carvalheiro, Ruan Veldtman, Awraris G. Shenkute, et al., "Natural and Within-Farmland Biodiversity Enhances Crop Productivity," *Ecology Letters* 14 (3) (2011): 251–259.

CHAPTER 5

1. Charles Owens and L. L. Farrar, "Electric Heating of Honey Bee Hives," Agricultural Research Service, United States Department of Agriculture (1967): 24.

CHAPTER 6

1. Khalil Hamdan, "Crystallization of Honey," *Bee World* 87 (4) (2010): 71–74, www.ibra.org.uk.
2. Paulus H. S. Kwakman, Anje A. te Velde, Leonie de Boer, Dave Speijer, Christina M. J. E. Vandenbroucke-Grauls, and Sebastian A. J. Zaat, "How Honey Kills Bacteria," *Journal of the Federation of American Societies for Experimental Biology* 24 (7) (1 July 2010): 2576–2582.
3. Eva Crane and Penelope Walker, "Composition of Honeys from Some Important Honey Sources," *Bee World* 65 (4) (1984): 167–174, www.ibra.org.uk.
4. Mayuri Tanaka, Yoshihiro Okamoto, Takashi Fukui, and Toshiyuki Masuzawa, "Suppression of Interleukin 17 Production by Brazilian Propolis in Mice with Collagen-Induced Arthritis," *Inflammopharmacology* (23 August 2011): 1–8.

CHAPTER 8

1. Martha Gilliam, Stephen Taber III, Brenda J. Lorenz, and Dorothy B. Prest, "Factors Affecting Development of Chalkbrood Disease in Colonies of Honey Bees, *Apis mellifera*, Fed Pollen Contaminated with *Ascosphaera apis*," *Journal of Invertebrate Pathology* 52 (1988): 314–325.
2. Marla Spivak and Gary S. Reuter, "Resistance to American Foulbrood Disease by Honey Bee Colonies *Apis Mellifera* Bred for Hygienic Behavior," *Apidologie* 32 (6) (2001): 555–565.
3. M. E. Nasr, G. W. Otis, and C. D. Scott-Dupree, "Resistance to *Acarapis Woodi* by Honey Bees (*Hymenoptera: Apidae*): Divergent Selection and Evaluation of Selection Progress," *Journal of Economic Entomology* 94 (2) (2001): 332–338.
4. Tibor Szabo, "Breeding Bees Resistant to Varroa Mites Is Like Trying to Breed Lambs That Are Resistant to Wolves," *The Buzz: Newsletter of the Iowa Honey Producers Association*, October 1993.
5. Lilia I. De Guzman, Thomas E. Rinderer, and Amanda M. Frake, "Growth of Varroa Destructor (Acari: Varroidae) Populations in Russian Honey Bee (*Hymenoptera: Apidae*) Colonies," *Annals of the Entomological Society of America* 100 (2) (2007): 187–195.
6. Dr. Frank Eischen and Sean Adams, "Smoking Out Bee Mites," USDA-ARS Honey Bee Research Laboratory, Weslaco, TX, www.ars.usda.gov/is/AR/archive/aug97/mitesmoke0897.htm.
7. Cédric Alaux, Jean-Luc Brunet, Claudia Dussaubat, et al., "Interactions Between Nosema Microspores and a Neonicotinoid Weaken Honeybees (*Apis Mellifera*)," *Environmental Microbiology* 12 (3) (March 2010): 774–782.
8. Paolo Tremolada, Marta Mazzoleni, Francesco Saliu, Mario Colombo, and Marco Vighi, "Field Trial for Evaluating the Effects on Honeybees of Corn Sown Using Cruiser® and Celest xl® Treated Seeds," *Bulletin of Environmental Contamination and Toxicology* (24 July 2010): 1–66.

9. Axel Decourtye, Eric Lacassie, and Minh-Hà H. Pham-Delègue, "Learning Performances of Honeybees (*Apis Mellifera L*) Are Differentially Affected by Imidacloprid According to the Season," *Pest Management Science* 59 (3) (March 2003): 269–278.

CHAPTER 9

1. David De Jong, "Africanized Honeybees in Brazil, Forty Years of Adaptation and Success," *Bee World* 77 (2) (1996): 67–70, www.ibra.org.uk.
2. J. S. Pettis, H. A. Higo, T. Pankiw, and M. L. Winston, "Queen Rearing Suppression in the Honey Bee—Evidence for a Fecundity Signal," *Insectes Sociaux* 44 (4) (1997): 311–322.
3. Jeffery S. Pettis, Anita M. Collins, Reg Wilbanks, and Mark F. Feldlaufer, "Effects of Coumaphos on Queen Rearing in the Honey Bee, *Apis Mellifera*," *Apidologie* 35 (6) (2004): 605–610.
4. C.A Mullin., M. Frazier, J.L. Frazier, S. Ashcraft, R.Simonds, et al.,"High Levels of Miticides and Agrochemicals in North American Apiaries: Implications for Honey Bee Health," PLoS ONE (2010), 5:e9754, www.plosbiology.org; H. M. Thompson, S. Wilkins, A. H. Batterby, R. J. Waite, and D. Wilkinson, "Modeling Long-Term Effects of IGRs on Honey Bee Colonies," *Pest Management Science* 63 (2007): 1081–1084.

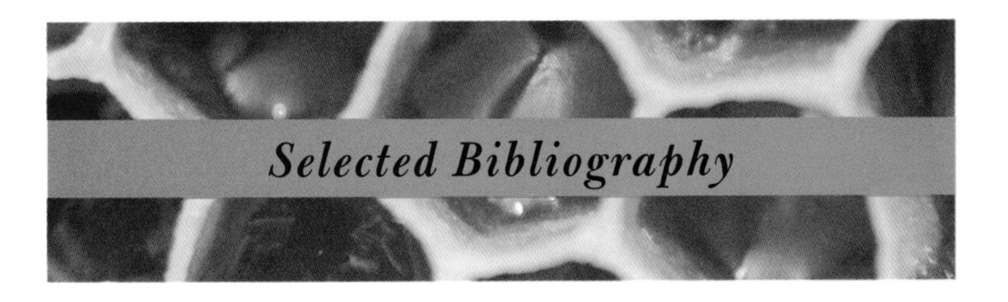

Selected Bibliography

Borror, Donald. *Insects*. New York: Houghton Mifflin Co., 1970.

Crane, Eva. *The Archaeology of Beekeeping*. London: Gerald Duckworth & Co. Ltd., 1983.

Dadant. *The Hive and the Honey Bee*. Carthage, Illinois: Dadant and Sons, 1998.

De Jong, David. "Africanized Honeybees in Brazil, Forty Years of Adaptation and Success." *Bee World* 77 (2) (1996): 67-70, www.ibra.org.uk.

Gould, James. *The Honey Bee*. New York: Scientific American Library, 1988.

Laidlaw, H.H. *Queen Rearing and Bee Breeding*. Kalamazoo, Michigan: Wicwas Press, 1997.

Munn, Pamela. *Varroa! Fight the Mite*. London: International Bee Research Association, 1997.

Prys-Jones, Oliver. *Bumblebees*. Slough, UK: Richmond Publishing Co. Ltd., 1987.

Punchihewa, R. W. K. *Beekeeping for Honey Production in Sri Lanka*. Sri Lanka Department of Agriculture, 1994.

Root, Amos Ives. *The ABC and XYZ of Bee Culture*. Medina, Ohio: A. I. Root Company, 2007.

Ruttner, F. *Queen Rearing: Biological Basis and Technical Instructions*. Bucharest, Romania: Apimundia Publishing House, 1983.

Sammataro, Diana. *The Beekeeper's Handbook*. Ithaca, New York, and London: Comstock Publishing Associates, 2011.

Seeley, Thomas. *Honeybee Ecology*. Princeton, New Jersey: Princeton University Press, 1985.

Tautz, Jürgen. *The Buzz about Bees*. Heidelberg, Germany: Springer-Verlag, 2008.

Traynor, Joe. *Almond Pollination Handbook*. Bakersfield, California: Kovak Books, 1993.

White, Elaine. *Super Formulas, Arts, and Crafts: How to Make More Than 360 Useful Products That Contain Honey and Beeswax*. Windsor, Ontario, Canada: Bench Mark Works, 1993.

Wilson, Edward O. *The Insect Societies*. Cambridge, Massachusetts: Harvard University Press, 1971.

Winston, Mark L. *The Biology of the Honey Bee*. Cambridge, Massachusetts: Harvard University Press, 1987.

BEEKEEPING PUBLICATIONS

American Bee Journal, published by Dadant & Sons, www.dadant.com

Bee Culture, published by the A.I. Root Co., www.beeculture.com

Beekeeping for Development, www.beesfordevelopment.org

Journal of Apicultural Research and Bee World, published by International Bee Research Association, www.ibra.org.uk

ONLINE RESOURCES

CiteULike: organizes academic papers, www.citeulike.org

For the Love of Bees: information about natural top-bar beekeeping, www.fortheloveofbees.com

International Bee Research Association: information on bee science and beekeeping worldwide, www.ibra.org.uk

Mendeley: a free reference manager and PDF organizer, www.mendeley.com

PLoS ONE: an inclusive, peer-reviewed, open-access resource from the Public Library of Science, www.plosone.org

ResearchGate: a network dedicated to science and research, www.researchgate.net

Index

Note: page numbers in *italics* refer to photographs and figures.

Chelsea Green Publishing is committed to preserving ancient forests and natural resources. We elected to print this title on paper containing at least 10% postconsumer recycled paper, processed chlorine-free. As a result, for this printing, we have saved:

10 Trees (40' tall and 6-8" diameter)
4,245 Gallons of Wastewater
4 million BTUs Total Energy
269 Pounds of Solid Waste
941 Pounds of Greenhouse Gases

Chelsea Green Publishing made this paper choice because we are a member of the Green Press Initiative, a nonprofit program dedicated to supporting authors, publishers, and suppliers in their efforts to reduce their use of fiber obtained from endangered forests. For more information, visit www.greenpressinitiative.org.

Environmental impact estimates were made using the Environmental Defense Paper Calculator. For more information visit: www.papercalculator.org.

About the Authors

LES CROWDER has devoted his entire adult life to the study and care of honeybees. Dedicated to finding organic and natural solutions for problems commonly treated with chemicals, he designed his own top-bar hives and set about discovering how to treat disease and genetic weaknesses through plant medicine and selective breeding. He has been a leader in his community, having served as New Mexico's Honeybee Inspector and president of the state's Beekeepers Association. He is an avid storyteller, and has spoken at the New Mexico Organic Farm Conference every year for over fifteen years.

Les built his own home, using the hexagonal shape in honeybee combs, and has made farming his career. He is also a certified teacher and enjoys teaching children Spanish and science. He raises sheep and poultry as well as honeybees.

He has three wonderful children, Emily, Peter, and Ben. Les is married to Heather Harrell, and they work their bees and certified organic/biodynamic farm together.

HEATHER HARRELL moved to New Mexico in 1996 from her home state of Vermont to pursue her masters degree in Eastern Classics, having long had an interest in the art of meditation and yoga and a yearning to pursue a career in academia. Her love of nature soon had her pursuing a life as an organic farmer, focusing on flowers, then medicinal herbs.

Over time, and through her work with honeybees, Heather has moved her focus to the study of multiuse permaculture plantings, which support a diverse network of interrelationships in the natural world. Along with a wide variety of vegetables, she grows medicinal herbs, which offer nectar and pollen to pollinator species. She is very interested in how soil biology is affected by using biodynamic methods of planting, and is currently studying compost teas incorporating various types of manures and plant materials.

Heather is known to make a good pie, and preserves more food each year. She has a wonderful daughter named Hannah, who helps out on the farm by bringing light and love into the garden.